Guide to Microturbines

Guide to Microturbines

Bernard F. Kolanowski, BSME

Routledge
Taylor & Francis Group

LONDON AND NEW YORK

Published 2020 by River Publishers
River Publishers
Alsbjergvej 10, 9260 Gistrup, Denmark
www.riverpublishers.com

Distributed exclusively by Routledge
4 Park Square, Milton Park, Abingdon, Oxon OX14 4RN
605 Third Avenue, New York, NY 10017, USA

Library of Congress Cataloging-in-Publication Data

Kolanowski, Bernard F.
    Guide to microturbines/Bernard F. Kolanowski
        p. cm.
    ISBN 978-8-7702-2354-6 (print) -- ISBN 978-8-7702-2234-1 (electronic)
    1. Gas-turbine power-plants. I. Title.

    TK1076.K66 2004
    621.1'99--dc22

                                                    2004046904

Guide to microturbines/Bernard F. Kolanowski.
First published by Fairmont Press in 2004.

Routledge is an imprint of the Taylor & Francis Group, an informa business

0-88173-418-7 (The Fairmont Press, Inc.)
978-8-7702-2354-6 (print)
978-8-7702-2234-1 (online)
978-1-0031-6854-6 (ebook master)

Dedication

I couldn't have written this book without the encouragement of Mr. Bill Payne, former editor of *Cogeneration and Competitive Power Journal*, and his recommendation to the publisher to see such a book published. I hope I've confirmed Bill's confidence in me.

AND

To my considerate wife, Mary Beth, who allowed me to take time away from her to accomplish this.

Table of Contents

Acknowledgments

The following companies and individuals have contributed to this book:

Mr. Robin Mackay
Bowman Power Systems
Capstone Turbine Corporation
Elliott Energy Systems
Ingersoll-Rand Power Works
Turbec AB
Toyota Turbine Co.
ALM Turbine Co.
Orkas Energy Endurance and Dr. Ake Almgren
Mr. Jim Clyde

Chapter 1

Introduction and History of The Microturbine

I WAS THERE at the birth of the microturbine. North American Co-generation was owned by Herb Ratch who knew Robin MacKay and Jim Noe, two engineers that had left Garrett Corporation after it had merged with Allied Signal. Robin and Jim decided to form a company to develop a small gas turbine that might be useful in the automotive market. This was in 1988 when NoMac Energy came into being.

Funding was important to NoMac and they decided to solicit a grant from the Gas Research Institute for those funds. Robin asked Herb to help prepare the proposal to the GRI and as I was then representing North American Cogeneration, they solicited my help in doing some of the writing. A prototype machine had already been manufactured by NoMac and I not only held many of the key component parts in my hands, I saw the first microturbine in operation at NoMac's facility. Little did I know that some 10 years later the microturbine would be such an instrumental part of my future.

It's believed the word "microturbine" evolved from the fact that it is a true gas turbine demonstrating all the characteristics of a gas turbine, but simply smaller in power output. No formal boundary exists as to when a gas turbine becomes a microturbine or vice-versa. However, it is generally accepted that zero to 300 kilowatts is the "range" of the microturbine. Pratt & Whitney gave this credence when they developed their 400 kilowatt unit and called it a "mini turbine!"

NoMac's company evolved into what is today the Capstone Turbine Company headquartered in Chatsworth, CA. While Capstone is rightly considered the originator of today's microturbine, it was not until December of 1998 when commercially available and reliable units were finally marketed. An elaborate article in the April 1, 1996 issue of Fortune Magazine prematurely touted the advent of this "pint-sized power house." False starts plagued Capstone, mostly in the power electronics area, before they finally solved the problems. After all, getting 60 hertz, AC current from a generator spinning at 96,000 rpm was no mean trick.

So, the microturbine is a recent development. The vast majority of gas turbines today are jet engines, turboprops or turboshaft engines. Renowned for their high power to weight ratio, extreme reliability and low maintenance, these engines dominate the aircraft industry. Derivatives of these turbines drive electric utility generators, power pipeline compressors and propel ships. A separate class of industrial gas turbines is used in power generation and other heavy duty applications. Almost all of these industrial gas turbines, however, are rated in the thousands or tens of thousands of kilowatts with more than a few over a hundred thousand kW. The microturbine has many big cousins, but it is a gas turbine extolling the same advantages as those brawny brutes.

The one noted advantage, however, is the ultra-low emissions that the microturbine emits. One disadvantage is that the small size of the compressor and turbine wheels limits the component efficiency, holds down the pressure ratio and prevents the turbine wheel from being internally cooled. Thus, the efficiency of a small, simple cycle gas turbine is well below that of a reciprocating engine—14% vs. 40%. Small production quantities have meant relatively high prices compared to the 100-year-old reciprocating engine that is installed in virtually every moving vehicle driving down the highway. These two factors have limited market penetration.

EVOLUTION

The primary application for small gas turbines has been in the aircraft industry. Most commercial and military airplanes use pneumatic starters to start their jet engines or turboprops. Most air conditioning on these aircraft is air cycle and requires a source of clean, oil free, compressed air. Simultaneously electric power is needed. Accordingly, gas turbines were developed that have over-size compressors that can be bled to provide the needed compressed air. At the same time the gas turbine drives an alternator through a reduction gearbox to provide electricity that is typically 400 hertz.

When these auxiliary power units, APUs, are installed in an aircraft, the low weight, low maintenance and high reliability overcome any concerns about high cost and fuel consumption, especially since the operating hours are few.

In the early 1960s The Garrett Corporation adapted two of their 85 series APUs to run on natural gas and drive 200 kW generator sets. Equipped with exhaust heat boilers, they were installed in one of the early gas turbine-driven cogeneration systems. With the concept proven, Garrett then developed the 831 series industrial gas turbines that were derived from their 331 series turboprop engines. Initially rated at 218 kW, the rating was eventually increased to 515 kW. Several hundred of these units were installed in a wide variety of cogeneration systems. Reliability was extremely high and systems were installed to provide precise power for the central computer systems of United, Continental and Western airlines as well as the United States Air Force Automated Data Weather System and also at a savings and loan institution.

The Boeing Company designed some of the first small gas turbines during World War II. This was an exercise to learn the characteristics of gas turbines when Boeing started the design of the B-47. Although this was a six engine jet propelled bomber, it was originally planned to be a turboprop. Hence, Boeing's small gas turbine had an output shaft. Indeed, this was the first free turbine engine. Separate turbine wheels drove the compressor and the output shaft. Thus, the output shaft speed could be varied all the way down to stall.

The U.S. Navy noticed that the Boeing 502s were primarily stainless steel and aluminum and thus had no magnetic signature. They purchased several hundred for mine sweepers. During the late 1950s and early 1960s, Boeing pioneered many gas turbine-powered applications such as anti-submarine drone helicopters, an oil cementing truck, a bulldozer, fire engines, a fire boat, high speed launches, army tanks and even an Indianapolis race car. Two were delivered for installation in a Thunderbird and a Fairlane. Many were used to provide compressed air to start 707s, DC-8s and other aircraft that did not have auxiliary power units.

The first cogeneration system powered solely by gas turbines was installed at Southern California Gas Company in their Downey, CA, facility in 1962. The two gas turbines were Boeing 502s rated at 140 kW each.

Boeing even developed a 100 horsepower outboard motor. It was a technical triumph. It weighed 100 pounds less than the 80 horsepower conventional outboards that were the largest in production in those days. It also burned less fuel. However, Boeing was not in the consumer products business and the outboard program was terminated and three years later Boeing sold their gas turbine business to Caterpillar

During the 1960s many automobile companies developed prototype gas turbines for automobiles. Rover was the first. Chrysler put fifty units out in the field for testing. General Motors and Allison built several different models. Ford drove a truck across the United States powered by their gas turbine. Daimler Benz, Volkswagen, BMW, Toyota, Nissan and others all built gas turbines. To solve the fuel consumption problem, automotive gas turbines used heat exchangers to pre-heat the air going into the combustion chamber using the normally wasted heat in the exhaust. These heat exchangers were sometimes recuperators but more often regenerators. Whereas a recuperator is a simple fixed boundary heat exchanger, a regenerator is a wheel that rotates through the exhaust picking up heat, and through the compressor discharge where it preheats the air going to the combustor.

Regenerators are compact and effective. However, they are usually ceramic and have a problem with cracking. They also have a sealing problem as the wheel rotates through both atmospheric pressure exhaust and high pressure compressor discharge air. Problems with the regenerator, cost of production and turbo lag killed the automobile gas turbine. Turbo lag was a particular challenge. To lower fuel consumption when an automobile was stopped in traffic or at a red light, the rpm was lowered. When the light changed to green, power was limited until the turbine spooled up which could take several seconds. You can imagine the cacophony of horn sounds behind a turbine powered vehicle.

The most interesting vehicular gas turbine in those days was the Ford 705. It was essentially a turbocharged, recuperated gas turbine. It achieved an efficiency of over 36%, which was better than the diesels being sold in those days. However, it was not really a small gas turbine in that the two versions were rated at 300 and 600 horsepower. Unfortunately, cost was a problem and Ford reverted to conventional regenerated gas turbines.

The survivor in small gas turbines was The Garrett Corporation Garrett was one of the Signal Companies. When Signal merged with Allied, Garrett became part of AlliedSignal Aerospace. When AlliedSignal bought Honeywell and adopted the Honeywell name, Garrett became part of Honeywell.

The two largest divisions of Garrett were the ArResearch Manufacturing Company of Arizona located in Phoenix, and the AiResearch Manufacturing Company of Los Angeles originally located in Los Angeles and later in Torrance. Phoenix was the dominant manufacturer of small gas turbines with many tens of thousands of units installed as auxiliary power units on board aircraft and in ground carts. The 831 series gas turbines discussed above were built by Phoenix. However, it was Torrance where the precursors to the modern microturbine were developed.

Torrance was by far the largest manufacturer of aerospace environmental control systems. They are used for air conditioning and heating commercial and military aircraft. Most of these units are air cycle and include high efficiency compressors and turbines mounted on fluid process bearings commonly known as air bearings. These bearings require no lubricants and no outside source of compressed air.

Torrance also built high-speed generator, high-speed refrigerant compressors, recuperators and gas turbine control systems. With support from the gas industry and the Department of Energy, Torrance used their expertise to develop prototypes of a radical new 10 ton heat pump in the 1970s. A 12 horsepower, natural gas fueled, recuperated gas turbine was used to drive a centrifugal refrigerant compressor which replaced the electrically driven compressor in a conventional heat pump. To keep the refrigeration system hermetic, the gas turbine drove the compressor through a magnetic coupling. To eliminate the need for a natural gas compressor and to enlarge the components so that they could be more efficient, a subatmospheric cycle was used.

In a conventional recuperated cycle gas turbine, outside air is compressed, preheated in the high pressure side of a recuperator, heated to higher temperature in a combustor and expanded through a turbine wheel, which powers both the compressor and the load. It then enters the low pressure side of the recuperator where heat is transferred to the high pressure side of the recuperator.

Subatmospheric cycles are different in that outside air enters the high pressure side of the recuperator first. The air is preheated in the recuperator, further heated in the combustor and expanded into a partial vacuum in the turbine, which drives both the compressor and the load. As in the conventional cycle, the air then enters the low pressure side of the recuperator where heat is transferred to the high pressure side of the recuperator. The air then enters the compressor, which sustains the partial vacuum before being discharged to atmosphere.

The conventional cycle and the subatmospheric cycle use essentially the same components. Both are Brayton cycles, the difference being where the air enters and leaves the cycle. The key advantage of the subatmospheric cycle is that the combustor is at, or very slightly below, atmospheric pressure. Thus, natural gas at normal delivery pressures will flow into it and there is no need for a fuel gas compressor, which is expensive and generally inefficient.

The second advantage is that the power output is reduced by a factor approximately equal to the pressure ratio. Thus, in a very small gas turbine the compressor and turbine are significantly larger and therefore have higher component efficiencies. In larger units, this becomes a disadvantage as the power available for a given piece of turbomachinery is reduced. Another way to look at subatmospheric cycles is that its performance is essentially the same as that of a conventional cycle that is operating at altitude.

Prototypes of the 10-ton heat pump achieved high levels of performance. They also demonstrated the feasibility of the subatmospheric cycle. The power unit was the first example of a gas turbine designed for production that ran on air bearings and required no lubrication and no source of compressed air. These bearings are ideally suited for high speed machinery where there are no gears because the load is driven at the same speed as the gas turbine. No power take-off is needed to drive an oil pump. Indeed, there is no oil pump, no oil sump, no oil cooler, no oil changes and no need to top off or check oil levels. Similarly, with no water cooling requirements, elimination of open or closed water cooling systems and the attendant treatment, pump and maintenance problems were also eliminated.

The prototypes proved the concept and demonstrated the performance. Unfortunately, government studies erroneously claimed that natural gas would be in short supply and that the price of natural gas would dramatically increase relative to the price of electricity. This would cripple the economics of gas fueled heat pumps and the program was discontinued. The fact that natural gas prices and electric prices followed the same curve in many parts of the country was mitigated when most public utilities were regulated and could not readily change their rates despite higher fuel costs. In this day of de-regulation of power companies, that factor would be less of a hindrance to marketing natural gas fueled technology.

Smaller, three ton heat pumps were also developed at Torrance for the residential market, but suffered a similar fate even though the concept included a bottoming cycle, which increased the output and the efficiency by about fifteen percent. Because the cycle was subatmospheric, the exhaust discharged out of the compressor. The exhaust was hot because of the heat of compression. Thus liquid refrigerant from the heat pump could be pumped up to pressure and vaporized in the gas turbine's exhaust. It could then be expanded through a turbine wheel that would be mounted on the back of the refrigerant compressor. The turbine wheel would then discharge the expanded refrigerant into the same condenser that the refrigerant compressor discharged into. As both wheels used the same refrigerant, small amounts of leakage did not matter.

The method of starting such a system was interesting. In a conventional gas turbine-driven generator set, the generator can be used as a starter motor. But there was no generator in this concept. However, there was liquid refrigerant and a refrigerant turbine wheel. Thus liquid refrigerant could be contained and heated until it vaporized using an electric resistance hater. This vaporized refrigerant could then be suddenly released to flow through the refrigerant turbine wheel causing it to spin up. As the refrigerant turbine was connected to the gas turbine through the magnetic coupling, the gas turbine would also spin up and would start.

When the energy crisis eased and it was realized that natural gas would be available, Torrance started on the design of the Advanced Energy System or AES. Basically, it was a recuperated gas turbine-driven generator set rated at 50 kW. It used a conventional rather than a subatmospheric cycle. The rotating group consisted of a permanent magnet with a compressor wheel mounted on one end and a turbine wheel mounted on the other end. The rotor group ran on air bearings so no lubrication was needed. Other than cooling fans and a fuel pump or natural gas compressor, this assembly was the only moving part in the system.

With an eye on fuel consumption, the AES had a recuperator. This heat exchanger transferred heat from the hot turbine exhaust to the compressed air entering the combustor. Thus the combustor needed less fuel to bring the air up to the required temperature. Fuel consumption was cut roughly in half compared with a gas turbine without a recuperator. However, there was still a lot of heat left in the exhaust. Making use of this energy for heating or cooling a building or for an industrial process such as drying could raise the system efficiency up into the 80% range.

Prototypes ran well. However, the Signal Companies merged with Allied and became AlliedSignal, as mentioned above. The Garrett divisions in Torrance and Phoenix became part of AlliedSignal Aerospace. The non-aerospace, non-military projects were terminated and the AES became an APU where it was installed in an army tank as a demonstration. Interestingly enough, AlliedSignal returned to this field several years later to develop the Parallon microturbine, the rights of which were sold to General Electric.

Robin MacKay had been instrumental in many of these programs. At Boeing, he initiated, sold and installed the cogeneration and oil field systems. He also worked on the outboard motor. At Garrett, he was responsible for most of the cogeneration sales and developed the concepts for the two subatmospheric gas turbine programs and the Advance Energy System. He wrote numerous papers and held several patents.

FULFILLMENT

MacKay took early retirement from what was now AlliedSignal and contacted Jim Noe who had been in engineering at Garrett, had worked with MacKay on several projects and held various patents on air conditioning and on subatmospheric gas turbines. Thus was the start of NoMac in 1988. Fortunately, AlliedSignal was gracious enough to grant NoMac licenses to some of the patents that had been issued to MacKay and Noe while they were at Garrett but were now assigned to AlliedSignal. The key patent licensed was the one for the residential heat pump that Garrett had designed but not built.

The company was very small. For the first five years it consisted of MacKay and Noe plus, intermittently, one engineer, one draftsman and one secretary. NoMac relied heavily on outside consultants for detailed design and analysis. NoMac also entered into joint venture with Tiernay Turbines called MTN Energy Systems. MTN stood for MacKay Tiernay Noe. Eventually, this joint venture was dissolved.

The original objective was to develop the residential heat pump with funding from the gas industry. The market for residential heat pumps and air conditioners was estimated to be in the six million units per year range. The projected coefficients of performance (COP) were 2.0 in the heating mode and 1.6 in the cooling mode. Thus, the units, if successful would have offered dramatic savings in both energy consumption and energy cost when compared with the best available units at the time.

A derivative version was also to be developed. The magnetic coupling and the refrigerant compressor and turbine were to be replaced with a generator. The objective was to build small generator sets in the three to six kilowatt range with the first applications aimed at the recreational vehicle market.

The gas industry was very enthusiastic about the potential of a very efficient gas fueled air conditioner. They were somewhat less enthusiastic about a very efficient gas fueled heating system that would only use half the gas that the best residential furnace then available used. Accordingly, the decision was made to increase the rating to 25 tons. The new markets were to be commercial establishments such as stores and factories where lots of air conditioning but very little heating would be needed.

With the increase in size, the subatmospheric cycle became less attractive and a conventional positive pressure cycle was used. An analysis was also made of the potential for an air cycle heat pump using principles developed for and commonly used on aircraft. Jim Noe had been one of the principal designers of these systems at Garrett.

One of the problems encountered was that the air cycle system optimizes at a lower speed than the gas turbine. This means that both systems should not share a common shaft. The answer was to use the gas turbine to drive a generator that was electrically connected to a motor driven, air cycle heat pump. By having a different number of poles on the generator and on the motor, the two units could operate at different speeds. The gas turbine-driven generator set evolved into what is now the Capstone 30 kW microturbine.

Various contracts for demonstration or study were received. They were from Southern California Edison for air cycle heat pumps, GRI for recuperators, NASA for recuperators, California Air Resources Board for low emission catalytic combustors and Ford for a generator set to keep batteries charged on an electric vehicle.

In 1993, venture capitalists invested in NoMac and the company took off. Renamed Capstone Turbine Corporation, it has shipped several thousand microturbines that have accumulated over three million hours of combined operation as of early 2003. Some of the early 1999 units have exceeded 35,000 operating hours without overhaul. They are currently rated at 30 kW and 60 kW with 200 kW in development. Typical applications include cogeneration, hybrid electric vehicles, precise power and distributed generation. Many are used in oil fields where they operate on casing head gas that would normally be flared. They are also used in landfills where they operate on the low Btu gases that are generated by the decomposing garbage. Sewage treatment plants using the anaerobic digesters give off methane gas in the 500—600 Btu/cubic foot range that can be burned in a microturbine with the waste heat put back into the digester in the form of hot water. Multiple units are common and the electronic controller can handle up to 200 units operating in parallel.

Today, Capstone is by far the dominant manufacturer of microturbines. While AlliedSignal's Parallon unit as marketed by Honeywell has been withdrawn from the marketplace, others such as Bowman Energy Systems, Ingersoll-Rand Power Works, Elliott Energy Systems and Turbec are currently manufacturing microturbines and their products as well as Capstone's are covered in subsequent chapters in this book.

Chapter 2

Design Goals and Achievements For the Microturbine

SINCE WHAT IS NOW the Capstone Turbine Corporation became the first marketable microturbine it is proper to know what goals were sought after when this turbine was developed for commercial operation.

The design team broke the components of the microturbine into three distinct areas:

1. Fuel System
2. The Engine
3. The Digital Power Controller

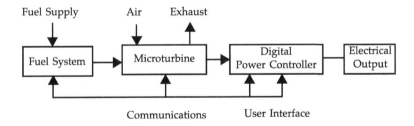

FUEL SYSTEM

Capstone wanted a versatile machine, able to burn a variety of fuels. They envisioned both gaseous and liquid fuels could be burned in this turbine, just as many different types of fuels were being burned in the larger gas turbines. Since the gas turbine is a mass flow machine vs. the reciprocating engines which are a volumetric machine, the designers realized that as fuel quality varied the mass of gas entering the combustion chamber had to vary. That meant the fuel system had to be designed to handle different masses of fuel flow.

Further, it was not just the "quality" gaseous fuels such as natural gas and propane they wanted to see burned, but also the low quality fuels coming out of land fills and digester plants, methane laden, but often accompanied by non-combustible gases such as carbon dioxide, and nitrogen. Therefore, for a gas turbine that was designed to burn 420,000 Btu/hour of a fuel type, and produce a nominal 30 kWh, the quantity of gas varied with the quality of the gas. For instance, natural gas is nominally rated at 1,000 Btu/cubic foot. To burn this gas, the fuel valve had to have the ability to pass 420 cubic feet per hour of fuel to the combustion chamber.

However, for a land fill gas having only 300-400 Btu/cubic foot heating values, the amount of gas that has to be introduced to the combustor is 1400 cubic feet per hour at 300 Btu/cubic foot and 1050 cubic feet per hour at 400 Btu/cubic foot. The unit still produces the nominal 30 kWh of electricity. Therefore, the fuel proportioning valve must be larger to handle this higher volume of gas entering the combustor.

Microturbines, like any gas turbine, require higher pressure fuel to the combustion chamber to match the pressure to which the supporting compressor pressure ratio produces. In the Capstone Model C-30 a nominal 55 psig fuel pressure is required when burning 1000 Btu/cf natural gas vs. as high as 70 psig fuel pressure when burning low Btu/cf gas as found in landfills and digesters. Any decrease in these fuel pressures will result in lower electric power output. Unlike its reciprocating brethren, gas turbine engines do not compress the air/fuel mixture in the combustion chamber but must rely on external compression to provide the proper combustion pressure. In the reciprocating engine the action of the piston in an enclosed volume provides the desired pressure ratio. Consequently, reciprocating engines can burn natural gas and low Btu gases with low inlet pressures and usually require no external compression.

Fuel cleanliness is also important in the gas turbine. Spinning at 96,000 rpm, the Capstone C-30 cannot tolerate moisture droplets in the fuel stream nor any solids. Any moisture contained in the fuel must be at least 18°F above its dew point anywhere within the fuel connections and the system between the microturbine fuel inlet and microturbine fuel manifold block.

Fuels coming from digesters, landfills gathering systems, oil field flare gases and the like may contain elements harmful to the turbine. While hydrogen sulfide is often present in such gases, the Capstone microturbine can tolerate up to 7% H_2S in the gas stream without fear of corrosion affects. Generally, H_2S is prevalent in oil field flare gases while the bio-gases, the proscribed term for digester and landfill gases, may have small percentages of H_2S while containing moisture and siloxanes.

The term, bio-gas, refers to the biological conversion of waste material to methane in an anaerobic digestion of organic waste materials. Anaerobic refers to a process that occurs in the absence of oxygen. Digestion refers to a biological process performed by microbes or bacteria (commonly known as bugs), which accomplishes the digestion of food. The bugs consume the organic waste material, rendering its solid residue essentially inert. The process occurs in the presence of water, ideally with the temperature and pH controlled to optimize the digestion reactions and the health of the bugs. The primary product is methane (CH_4), accompanied by carbon dioxide (CO_2). Typically the gas ratio is 60:40, 60% methane to 40% carbon dioxide.

Siloxanes may be indigenous to these bio-gases. Siloxanes are compounds containing the structural unit R2SiO where R is an organic group or hydrogen and SiO is silicon oxide. Not all bio-gases contain siloxanes as their presence depends on the material that is decomposing on whether silicon is present. Siloxanes are used extensively in consumer products to act as a volatile dispersant agents to help evenly spread organic based specialty chemicals. Deodorant, lipstick and makeup, plus many other products use siloxanes. As man-made compounds that typically are washed down the drain or thrown in the trash, siloxanes are always found in landfill and wastewater treatment plant digester gas. Not all digester plants contain these throw away items. Digesters are finding common use in dairy and hog farms as manure digesters, breweries and ethanol plants and food processing plants. When siloxanes are present in the gas stream it behooves the installation of microturbines to remove such contaminants. Otherwise, the silicon, under high temperature, forms a glass like coating on the components of the microturbines power wheel blades and entrance to the recuperator. This coating will limit the power output of the turbine as well as its life.

Siloxane removal equipment consists of the adsorption of the siloxanes in the pores of graphite media. One such filter is manufactured by Applied Filter Technology (AFT) and can respond to the gas analysis with different pore sizes and structures as well as layering them differently for each biogas to optimize overall performance.

Figure 2-1 shows a typical flowsheet for biogas fuel preparation.

Capstone wants to see a limitation of 5 ppb or 5 parts per billion by volume for siloxanes. Since this is the detectable limit for siloxanes it means the fuel must contain no detectable level of siloxane. When detected, filtration equipment must be installed in the fuel stream ahead of the turbine.

Liquid Fuels

The Capstone microturbine can also burn liquid fuels, specifically diesel and kerosene. However, other liquids have been tested such as fish oil and other more "exotic" liquids from various processes. Heating value, purity, viscosity and moisture are the determining factors as to whether the more exotic fuels can be burned.

For diesel fuel the calorific value in the HHV table is 19,900 to 20,000 Btu/pound. Kerosene is the same. Viscosity ranges from 1.3 to 4.1 centistokes for diesel and 1.0 to 1.9 centistokes for kerosene. Diesel fuels approved are ASTM D975 No. 2-D and 1-D as well as their counterparts in low grade sulfur content. Kerosene is

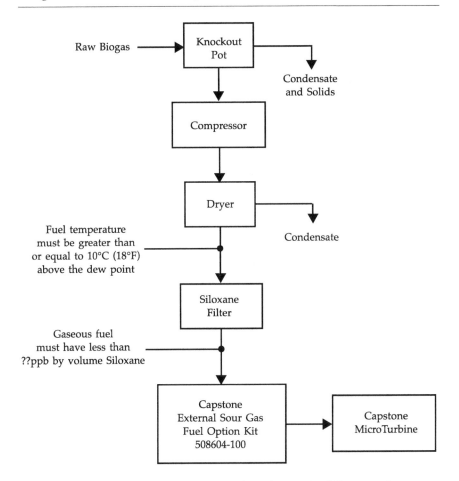

Figure 2-1. Typical Flowsheet for Biogas Fuel Preparation

per ASTM D3699 1-K and JIS K2209.

To be considered as a possible fuel for the microturbine, liquid fuel properties should fall into the specifications shown in Table 2-1. Table 2-2 shows limitations for liquid fuel contaminants.

When considering new fuels to be burned in the microturbine Capstone evaluates three components:

1. Combustion System Evaluation Concerns
2. Fuel System Evaluation Concerns
3. Microturbine (Engine) Evaluation Concerns

Table 2-1.

	Maximum	Minimum
Kinematic Viscosity-centistokes	14	1
Specific Gravity @ 68°F	0.95	0.75
Cloud Point -°F	-18	—
Pour Point -°F	-18	—
Flash Point -°F	150	100
Calorific Value—Btu/lbm	20,000	15,000
Vapor Pressure—psia	3	0
Initial Distillation Point -°F	350	250
Final Distillation Point -°F	700	—

Table 2-2.

	Maximum	Minimum
Free water @ 68°F- % mass	0.05	0
Particulate Size—microns	2.0	0
Particulate Quantity—ppm mass	5.0	0
Sulfur—ppm mass	10,000	0
Chlorine—ppm mass	1,500	0
Fluorine—ppm mass	150	0
Ash—ppm mass	100	0
Sodium + Potassium—ppm mass	0.5	0
Vanadium, Calcium, Lead & all Other contaminants—ppm mass	0.5	0

Combustion System Concerns include the fuel type that may affect the stability of the combustion system. Fuels highly diluted with inert items, as well as different liquid fuels, may have degraded stability at low power levels. Liquid fuels especially in the atomization ability has a significant effect on stability Fuels containing significant amounts of hydrogen and acetylene may result in flashback and combustion system damage.

Combustor life may be affected when fuels with significant amounts of carbon monoxide and acetylene are burned. High combustor temperatures are the result. Liquid fuels with poor atomization qualities will also result in reduced combustor life.

Some fuels have inherently higher emissions levels. Liquids will generally have higher NO_x levels than natural gas. Since temperature of combustion and NO_x production are synonymous, fuels containing significant levels of carbon monoxide and acetylene will produce more NO_x. Atomization ability of liquid fuels will also affect NO_x production. Vapor lock with liquid fuels containing high vapor pressures will occur in the fuel injectors resulting in rough running and nuisance shutdowns.

Fuel System Concerns include the gaseous fuel temperature which may raise concerns regarding liquid formation, high vapor pressure, high density and high viscosity. Damage to the materials of construction may also occur with high or low temperature extremes of the fuel.

Fuel contaminants, additives, as well as some hydrocarbons may corrode or attack critical fuel system components. Fuels containing water and high hydrogen sulfide amounts may lead to even more severe corrosion effects. Condensed water or ice will result in fuel system malfunction causing eventual shutdown. Gaseous fuels temperature should be maintained at 18°F above the dew point of water throughout the fuel system.

Fuel valve limitations affect what fuels can be burned as discussed previously in this book. Extremely reduced volumetric heating values may not allow full power operation due to flow control valve limits. Conversely, fuels with too high a heating value may not be adequately controlled at low power operation.

Capstone incorporates a Rotary Flow Compressor in its Model C-30 which boosts the inlet gas pressure to required combustion chamber pressure. Any of the excesses listed above may have an adverse effect on this component causing premature failure. Too low an inlet pressure may cause air leaking into the gas fuel supply and run the consequent risk of an explosive mixture.

Reduced bearing life in the ball bearing RFC will also occur with too low an inlet pressure. Since Capstone has introduced the foil bearing RFC, some of the above concerns no longer exist, but air leakage should still be avoided in the foil bearing RFC as well as any external fuel gas booster compressors.

With liquid fuels, the deviation of the liquid fuel density from the parameters noted in the above table may result in incorrect metering of the liquid fuel, and/or other fuel control issues. High fuel viscosity may result in increased heating of the fuel pump internal components affecting performance and cold start problems due to its effect on atomization will also occur. Lower fuel viscosity may result in increased internal leakage of the fuel pump which may adversely affect the fuel metering accuracy, especially under low flow conditions.

Further, liquid fuel pump life may be adversely affected by the fuel viscosity, fuel lubricity, and fuel temperature in addition to any contaminants in the fuel that are outside of the requirements detailed within this chapter. Water is equally concerned with freezing causing fuel blockage and shut down as well as biological growth in fuel system components, especially the fuel filter.

Microturbine Engine Concerns deal with damage to the hot end components caused by excess fuel contaminants such as sulfur. Liquids found in gaseous fuels can cause turbine hardware damage. Surge may occur when attempting to burn fuels with low heating values. Surge is capable of destroying the engine components.

It is notable to observe that the concerns for combustion system and fuel system far outweigh the concerns for the engine itself. This is a tribute to the ruggedness of a gas turbine in general and that of the microturbine especially.

THE ENGINE

The typical microturbine engine consists of the rotor, the combustion chamber and the recuperator. The rotor contains the permanent magnet generator which is surrounded by the stator. Also mounted on the rotor is the compressor wheel and the turbine power wheel. In the case of many of the microturbines the rotor is supported by air foil bearings which require no liquid lubrication. An air foil thrust bearing is also used. In the case of the Capstone microturbine the components are as shown in Figure 2-2.

The one piece rotor as shown in Figure 2-3. is about 18" long and contains all the dynamic parts of the microturbine—the permanent magnet generator on the right, the compressor wheel in the center and the power turbine wheel shown on the left. It rotates at 96,000 rpm at full speed and kilowatt production.

The foil/air bearing is so called because it has air foil shaped segments that lift as air is introduced into the microturbine upon start up rotation. The shaft is suspended in space by an air cushion created by the foils. The only time the shaft comes into contact with the air bearings is when the unit is at rest. Upon shutdown the rotor is "grabbed" electrically to prevent coast down of the rotor thereby preventing any rubbing of the shaft on the air bear-

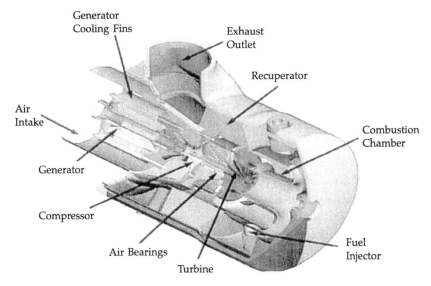

Figure 2-2. Capstone Microturbine

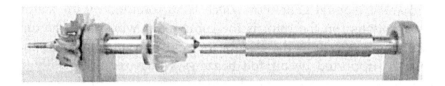

Figure 2-3. One Piece Rotor

- Reliability
- Fewer parts
- No liquid lube
- No liquid capacity
- High speed operation
- Low operating temperature
- No scheduled maintenance
- Load capacity increases as speed increases
- High temperature capability

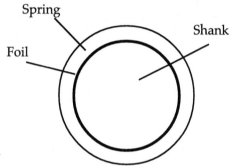

Figure 2-4. Foil/Air Bearing

ings. The obvious advantages of this construction is the elimination of an oil system and its associated sumps, pumps, piping, filters and the necessity of frequent oil changes.

Atmospheric air is introduced to the microturbine upon start up, which is accomplished electrically either by the connection to the grid or via batteries. Air enters to the left of the unit as shown in Figure 2-2 and goes across the generator portion of the rotor, cooling it as it passes, and into the compressor wheel. The air is compressed to a higher pressure before entering the recuperator and subsequently, the combustion chamber. The recuperator is a heat exchanger that allows the compressed air to be heated by the products of combustion thereby picking up significant efficiency points.

As the now superheated air enters the combustion chamber, fuel is introduced at the same pressure as the compressed air—55 to 75 psig depending on the size of the unit and the type of fuel being introduced—ignited via an igniter and the subsequent combustion introduced high temperature and high pressure gases to the power wheel. The power wheel supplies the power to produce the electricity from the generator and the energy needed to compress the air.

The hot gases leave the power wheel at elevated temperatures—1100°F—and pass through the other side of the recuperator, cooling the exhaust gases while heating the compressed air. The exhaust gases exit the turbine at approximately 600-700°F and at basically atmospheric pressure. For optimum performance the microturbine exit pressure should not exceed 8″ water column.

At a full power running speed of 96,000 rpm, the microturbine's production of 1600 hertz AC power requires the conversion of that frequency to usable 50/60 hertz AC. That accomplishment is discussed in the next section.

Figure 2-5. Recuperator type heat exchanger

ELECTRIC OUTPUT—
THE DIGITAL POWER CONTROLLER

In 1996 and the early part of 1997 Capstone believed the design was ready for commercial launch. The "beta" unit was called "Charlie" and about 100 units were sold and placed in operation.

Economic pressure from investors as well as the exuberance of the then CEO caused a target date of December, 1997 as the launch and it would be announced at the Power Gen conference in Dallas, Texas during that month.

During the early part of 1998 it became evident that there were too many problems with the design. It was decided to cancel the Charlie program and to buy back most of the units. The problems were not with the engine or the fuel system, but with the power electronics. Some blamed the problems on the fact that the early "Alpha" units were designed for 25 kWh output while the Charlie units had grown to 30 kWh output without any change in the power electronics system. Some units were left in the field to continue to test out the power electronics system with new components. This was called the B3 program. Six months later it was agreed to shut down all units.

However, the B3 program continued in house recognizing the symptoms that were causing the problems were from outside interference. Transients from the grid would cause the units to just stop. The sensitivity of the system to such transients as well as lightning storms indicated the system was not robust enough to ignore such transients. The B3 program added more control logic, improved software and generally beefed up the hardware to recognize the new power output. The new system was more robust concluding in the current system of a pure sine wave of power output, total ignorance of utility or outside influenced transients which resulted in the product today—a totally reliable power electronics system.

THE POWER ELECTRONICS:

There are two basic types of microturbines as related to power electronics modes. The one discussed so far is the single rotor type where the generator is a permanent magnet incorporated into the single rotor as shown in Figure 2-3. It is that rotor configuration that requires the power electronics system to derive 50/60 hertz AC current output. At 96,000 rpm the 1600 hertz frequency needs to be toned down. First, that AC power in rectified to DC power and then inverted through a Digital Power Controller to the 60 (or 50) hertz AC output.

The other type microturbine is one in which the power wheel is on a separate shaft. This configuration is often referred to as a split shaft turbine. The power wheel transfers its output to a conventional two or four pole generator via a gear reducer at 3600 or

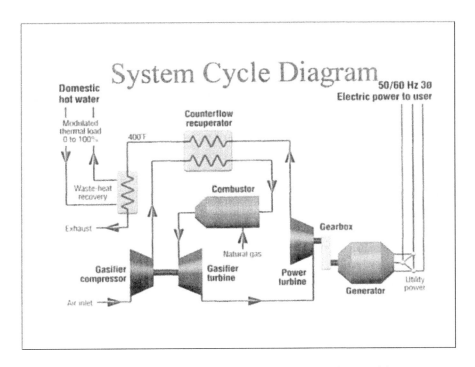

Figure 2-6. Ingersoll-Rand Power Works Microturbine

1800 rpm. It is this type of generator that is used in reciprocating engine-generator systems and while it is proven robust technology, it will need synchronizing equipment and discrete protection relays in order to be connected to the electric grid. Whether the gear reducer will prove to be a high maintenance item, along with its supporting lubrication system, remains to be seen by compiling operating hours in the field.

As opposed to the split shaft design, the use of an inverter has several advantages. Using a bi-directional inverter the generator can be motored at the start up. The inverter eliminates the mechanical gearbox and the maintenance related to it. The power electronics and the microprocessors used for control can also be used to provide most of the protection relay functionality. By taking voltage and frequency from the electric grid the synchronization is by definition automatic. Thus the need for separate synchronization equipment as well as the need for discrete relays for voltage and frequency protection is eliminated. The power electronics can be used for power factor correction and in many other aspects provide flexibility impossible with the other concept. One disadvantage of the power electronics is conversion losses and reduced robustness. However, as discussed in Capstone's "false" start, the robustness of the power electronics continues to improve and with those improvements, this disadvantage will go away.

The automatic synchronization of the power electronics design also lends itself more directly to the feature of Stand-alone capability. In other words, the full time power generator can also become a standby generator in the event of an electric grid power outage. This is often called dual mode operation.

Chapter 3

Dual Mode Operation

TO BE ABLE TO OPERATE in the stand-alone mode, the microturbine system will need a voltage and frequency source. Generally, an electric battery provides this source. Since the battery shall provide energy for the start up as well as handle power transients, the battery is oversized and used to only 20% of its capacity. The transient capability is needed since in the stand-alone operation the microturbine must follow the power demand of the load. This power demand can change faster than the microturbine can respond. The battery is used to bridge the power demand change with the response time of the turbine to ramp up or down.

Before understanding the full import of dual mode operation it is incumbent to know the working of the grid connected system. The microturbine is designed primarily to safely produce power in parallel with an electric utility. Protective relay functions inside the microturbine system ensure that power production is possible only if the utility waveform meets preset standards. A start command is given to the microturbine and this causes power to be drawn from the utility which starts the turbine engine. The system then outputs power per the power level setting, synchronized to the utility waveform.

The microturbine can be controlled to follow the demand load by instituting a power meter device in the control logic. This power meter senses system power needs and causes the turbine to slow down to meet those needs without exporting power to the grid. Not only does exported power bring little or no economic benefits, most utilities impose restrictions on exporting power. The microturbine accepts meter data signals for forward power flow, a + pulse, and reverse power flow, a – pulse at a rate proportional to the power flow at the control point where the meter is attached. Using these signals the microturbine will dynamically adjust its output power level to insure that no power flows back to the utility. This application is called "Load Following."

Power meters with KYZ outputs are required and are commercially available from such suppliers as ABB, Schlumberger, GE and Siemens.

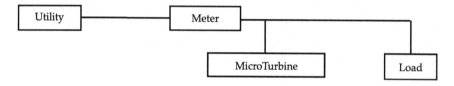

Figure 3-1. Typical Load Following Schematic

In stand-alone mode, the microturbine solely supports the load, providing required voltage, active and reactive power. Stand-alone capable microturbines are equipped with a battery and battery controller. The battery is used for both starting the microturbine and supplying transient energy to connected loads.

Often, it is necessary to power some loads prior to the microturbine reaching the capability to power external loads. Typically, fuel booster compressors are one such load. Without the proper fuel pressure to the combustion chamber, the microturbine will not function and with 55 to 90 psig pressures required, the local gas supplier is loathe to supply that high pressure into the customer's building. Hence, an auxiliary load controller (ALC) manages segregation of the loads to first start the fuel gas booster compressor and then the microturbine.

The combination of grid connect and stand-alone modes is called dual mode operation. This term refers to a microturbine's ability to operate both in parallel with the utility or isolated from the utility in stand-alone mode. Manual transfer between these modes of operation may be accomplished with a manual switch. However, since transition from one to the other configuration is most beneficial when this is done automatically, an automatic dual

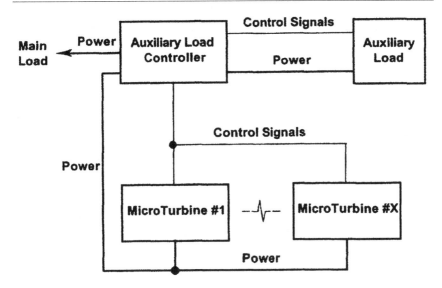

Figure 3-2. Stand-alone Mode showing ALC Device

mode controller (ADMC) is necessary.

The ADMC serves as a transfer switch and as an intertie disconnect between the grid and the load. Any load downstream of the ADMC is termed the "critical load." It begs logic to know that most microturbine installations, just like any standby generator, is usually not sized to handle the full load of the facility. Therefore, dedicated or critical loads are identified and supported by the standby generator, whether that generator is truly a standby unit or a full time power producing generator now operating in stand-alone mode.

In grid connect mode, when a grid outage occurs, the ADMC circuitry senses the outage and opens a motorized switch, isolating the microturbine and load from the utility.

Since the electric utility wants assurance that a total disconnect has occurred from its now "dead" lines, a time delay is necessary when the ADMC senses grid outage and before the microturbine can come on line under battery influenced power. This time delay may be anywhere from 30 seconds to 30 minutes. Therefore, an Uninterruptible Power Supply (UPS) system may be needed to support loads that cannot stand this time interruption, such as computer systems or safety lighting.

Figure **3-3** shows the complete system when an ADMC is involved. The ADMC senses grid power outage, the ALC starts the auxiliary load (fuel gas booster compressor) and the ADMC switches over to battery power to start the microturbine. Note that multiple microturbines can be controlled from this system.

When multiple turbines are connected to a system, a Power Server device can monitor load and cause matching of the power demand with the power output from the bank of turbines. This device can slow the turbines down in unison, or cause one or more turbines to be shut down completely and re-started when power demand rises. Up to 100 microturbines can be controlled in this fashion.

When grid power is restored, the ADMC senses grid power, and reverses its operation by shutting down the turbines and then restarting them in grid connect mode. While the shut down to stand-alone mode time is controlled more by the utility than the microturbine technology, the re-establishment of turbine power to grid connect mode is being shortened with the goal of seamless transition. Of course, all of this is predicated on the turbines being "hot" during the transition.

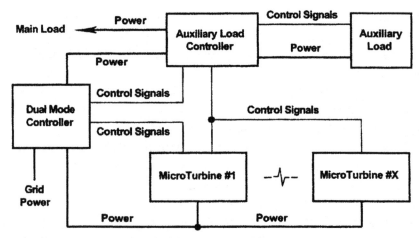

Figure 3-3. Typical Automatic Dual Mode Control Schematic

Chapter 4

Operational and Environmental Characteristics

A MICROTURBINE IS first a gas turbine and therefore follows the laws of physics where its performance is concerned. Being a mass flow machine as opposed to reciprocating engines which are volumetric machines, the gas turbine is more susceptible to changes in air density than the reciprocating engine. The gas turbine industry has set the inlet air temperature rating at 59°F for determining performance. Therefore, when a microturbine is rated at 60 kWh output it is at 59°F inlet air temperature. Variations from that temperature will cause changes in the power output of the turbine. Figure 4-1 shows the affect higher inlet air temperatures have on kilowatt output.

Note that the Efficiency curve is the top line while power output is the bottom line. In this case the Capstone Model C-60 microturbine will continue to put out 60 kWh up to 80°F, but the efficiency starts to drop after the 59°F rating point. At 100°F the power output of this turbine is at approximately 54 kWh. Below 59°F one would expect the power output to increase beyond 60 kWh since the air is more dense. However, the power electronics within the turbine prevent more power from being generated. In jet engines, where power is measured in thrust, the denser the air the more thrust (power) is produced.

Elevation Considerations

Elevation also affects turbine performance since the higher

31

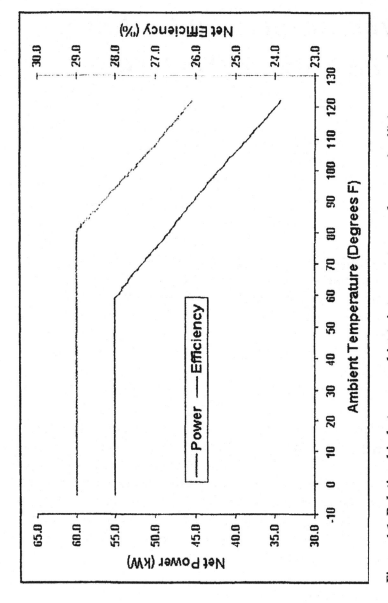

Figure 4-1. Relationship between ambient air temperature and power & efficiency production in a 60 kilowatt rated microturbine.

the elevation the less dense the air and the same de-rating phenomenon takes place. A rule of thumb is that for every 1000 feet above sea level, performance loss is 3% of full load rating. This elevation de-rating is often offset by the cooler average temperatures at the higher altitude. However, both inlet air temperature and elevation considerations must be accounted for when determining the power output at any given condition.

Back Pressure Considerations

Since stationary microturbines may be connected to exhaust stacks, heat recovery equipment or duct work for directing the exhaust gases to a drying furnace, it is important to know the relationship of power output vs. back pressure on the turbine exhaust. Figure 4-2 shows the decay in power and efficiency as the back pressure, measured in inches of water column, increases. Power output is maintained up to almost 7" water column and then decreases while efficiency starts minor decreases almost immediately. Capstone puts a maximum of 8" water column back pressure limit to prevent damage to their turbine.

For sizing ducts that will be attached to the exhaust of the microturbine various factors go into calculating system backpressure, including

- Anticipated length of the duct run
- The diameter of the duct
- Number of bends in the duct
- Type of material used which determines roughness and friction factors.

The Darcy Equation can be used to estimate friction loss as a function of duct diameter, duct length and exhaust velocity:

$$Hf = (f \, L/D + C)(V/4{,}005)^2$$

Where Hf = head loss due to friction in water
f = non-dimensional friction coefficient (0.2 for clean round metal ducting)

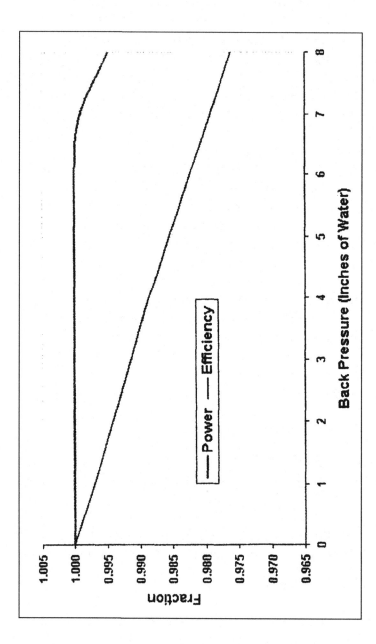

Figure 4-2. Affect of Exhaust Back Pressure on Turbine Performance

L = length of duct, ft.

D = diameter of duct, ft.

C = dynamic loss constant (0.16 for each 90° elbow; 0.57 for each backflow damper

V = velocity of exhaust, ft/min.

Starting and Stopping Considerations

Most microturbines can start and stop relatively fast. The time from a start command until power starts being delivered is about 30 seconds. Full power is reached in a couple of minutes. During the first phase of the start-up the microturbine consumes power for motoring the turbine.

Stopping the microturbine generally takes a couple of minutes from the stop command until completely stopped. The time is required for a controlled cool down of the engine, since a significant amount of thermal energy is "stored" in the recuperator and must be removed. Emergency stops as opposed to normal stops, require shorter sequences. Since the turbine is operating on air bearings, it is not prudent to allow the turbine to come to a rolling stop. This would place undue contact of the shaft with the air foil bearing causing premature wear. Therefore, the rotor is stopped electrically before that rubbing contact can be made.

Environmental Considerations

There are two categories of emissions. The first category is air quality related emissions which include pollutants such as NO_x (nitrogen oxide); SO_2 (sulfur dioxide); CO (carbon monoxide; particulate matter (PM); and volatile organic compounds (VOC).

The other category is emissions related to global warming, i.e. so called greenhouse gases, which include CO_2 (carbon dioxide), CH_4 (methane), N_2O (nitrous oxide) and the very powerful greenhouse gases that are not naturally occurring such as SF6 (sulfur hexafluoride).

Microturbines generally have very favorable air quality related emission characteristics thanks mainly to the fundamentals of lean-burn concept and relatively low turbine exhaust temperatures which helps control NO_x production. These very low emissions eliminate the need of any catalytic exhaust gas cleaning. This technology has made the microturbine exempt from air quality permitting in most, if not all, Air Quality Management Districts in the United States.

Natural gas fired reciprocating engines usually need catalytic exhaust gas cleaning to meet stringent emission standards. Diesel reciprocating engines need, in addition to catalytic exhaust gas cleaning, particle traps, but may still come short of meeting the most rigorous emission standards.

Table 4-1. Example of NO_x emissions for stationary power generators.

Equipment	NO_x lb/MWh	Comment
Diesel reciprocating engine	3.0-29.5	Catalytic exhaust cleaning
Natural gas reciprocating engine	1.6-26	Catalytic exhaust cleaning
Gas turbine	'0.3-3.8	Combined cycle (CCGT)
Microturbine, natural gas	<0.5	Capstone and Ingersoll-Rand
Microturbine, diesel	<.22	Capstone

Engines operating on fossil fuel generate greenhouse gases such as CO_2. The key to minimizing the CO_2 emissions is the efficiency of the engine. Microturbines by themselves are less efficient than reciprocating engines—approximately 30% vs. 40% respectively. However, in Combined Heat & Power applications this disparity in efficiency is mitigated since the lower efficiency of the microturbine will generate more usable heat than the reciprocating engine. Therefore, the overall amount of CO_2 production is virtually equal between the two types of engines.

Methane, CH_4, is over 20 times more negative than CO_2 as a greenhouse gas. Consequently, capturing methane from landfills, digesters in waste water treatment plants, oil field flares, agricultural decomposition, etc., for burning in both microturbines and reciprocating engines is one method of reducing methane into the atmosphere. Microturbines appear to be more flexible in burning the lower quality, contaminated methane coming from those sources than reciprocating engines. The Capstone turbine's ability to burn up to 7% by volume of H_2S (hydrogen sulfide) without deleterious affect is one example of that flexibility since H_2S is often found in many of those sources.

The guaranteed emissions levels for a Capstone microturbine operating at full power are shown in Table 4-2.

Part load operation will increase these emission levels, but the Capstone turbine has the ability to remain under 9 ppm NO_x levels, and lower, down to about 30% of full load power.

Noise Considerations

The primary noise sources from most microturbines are from the engine/generator assembly. Secondary noise sources are the electronics cooling fans, inductors in the Digital Power Controller. Vortices generated by the airflow create acoustic pressure fluctuations. Turbulence noise is generally broadband in nature and is typically most significant in the 1,000 to 5,000 Hertz octave bands in small gas turbine engines.

Other sources of noise internal to the microturbine are blade passing tones generated by compressor and turbine blades and electro-mechanical noise created by the interaction of the magnetic fields in the generator which is then transmitted through the stator into the stator housing. This latter noise is tonal in nature and is strongest at one and two times the rotational shaft speed—1600 hertz and 3200 hertz in the case of the Capstone microturbines.

Combustion noise is most prevalent in the simple cycle (nonrecuperated) engine near the exhaust side of the engine and is usually in the 400-800 hertz range. While additional noise can be created by structural vibrations due to the excitation of the natural frequencies of the engine and engine supporting structures, the air bearing support in microturbines provide a very soft bearing support so that vibration transmittal into the engine structure is minimal.

Table 4-2. Typical Emissions Levels for Capstone Microturbines

Component	Emissions (Max)	Emissions - Raw	Emissions				
	(ppmV @ 15% O_2)	(ppmV @ 18.3% O_2)	[lb/kW/hr]	[lb/hp-hr]	[gm/kW-hr]	[gm/GJ]	[gm/hp-hr]
HC	9	3.96	1.77E-04	1.32E-04	0.080	22.29	0.060
CO	40	17.63	13.73E.04	10.24$.04	0.624	173.33	0.467
NO_x	9	3.96	5.07E-04	3.79E-04	0.231	64.07	0.172
NO_x+HC			6.84E-04	5.10E-04	0.311	86.36	0.232
CO_2			1.69E+00	1.26E+00	766.0	2.13E+05	5.72.0
O_2			1.60E+01	1.19E+01	7270.0	2.02E+06	5420.0

It is through the acoustic paths that noise is transmitted from its source to the outside of the microturbine package. The three primary acoustic paths are inlet noise, exhaust noise and case radiated noise. Capstone has designed and installed inlet hoods to help attenuate inlet noise and commercially available exhaust silencers can attenuate exhaust noise. As stated above, case radiated noise is minimal in the Capstone, air bearing supported turbine.

Noise measurement made on Capstone units shows an average of 65 dba or less on a 10 meter radius. The model C-30 has a high frequency pitch at the very point that is most sensitive to the human ear, while the C-60 unit has no such frequency anomaly and is easily incorporated into hospitality sensitive sites.

Chapter 5

Markets for the Microturbine

THE MICROTURBINE has been earmarked as the logical generator for distributed generation due to its modularity, compact design, low emissions, no vibration, low noise levels and very low maintenance.

DISTRIBUTED GENERATION

If you are in the electric power industry as a utility, equipment supplier or a user of electricity you have heard the words—"Distributed Generation." What exactly does that mean? A good definition is, "...the integrated or stand-alone use of small, modular electric generators close to the point of consumption."

That's fine, but is it viable? In other words what do I as a consumer of electricity get if my electricity comes from a distributed generation module? The integrated part of the definition usually means the generator is interconnected with the local utility. The stand-alone part of that definition means that I may not be interconnected with the utility, or, if I am, I may be able to still have electricity if the utility power should fail. This latter aspect does give me value in that I will still be able to operate while my neighbors and competitors may not.

The U.S. electric power system is among the most dependable in the world achieving over 99% reliability. However, that number can be challenged in the future as deregulation gains a greater foothold. In a regulated industry the provider is rewarded with better rates as he spends capital to improve his service. A deregulated company may not be so prudent especially if he feels there is no competition to drive his customers elsewhere.

Furthermore, reliability is not just of the generating system, but also the delivery system. There is a real concern that the distribution lines feeding our homes and factories are deteriorating faster than they are being upgraded. Reliability to one person is of marginal concern while to others it is critical. The reliability jargon speaks in terms of "nines." Ninety-nine percent reliability is two nines. What this means is that system that is 99% reliable suffers 3.7 days/year of down time. To carry that further, three "nines" means a downtime of 9 hours/year; four nines—53 minutes a year; five nines—5 minutes/year; six nines—32 seconds per year and seven nines (99.99999%)—3 seconds/year.

The cost of an outage to a homestead is not nearly as critical as to an outage to some companies that depend on usage to earn their income. For instance the following shows how some types of businesses value their power:

Cellular Phone companies—	$41,000/hour of down time
Telephone Ticket Sales —	$72,000/hour
Airline Reservations —	$90,000/hour
Credit Card Operations —	$2,580,000/hour
Brokerage Operations —	$6,480,000/hour

Obviously, most of these types of companies have planned back up power to keep their critical needs operating in the event of a power outage.

So, reliability is the value proposition of the stand-alone capability of distributed generation. And, Value Proposition is the byword to any cost of implementing distributed generation. Another value to DG might be a savings in the cost of running transmission and distribution (T & D) lines to your project if it is not currently served by a utility. With real estate becoming more scarce in the normal utility service areas your project, be it a housing development, a new production factory, a golf course, an amusement park or a shopping arcade, the cost of T & D may be so exorbitant that an on site generation system may be the best value. In that case not only would the T & D expenses be deferred and eliminated, but so would the ongoing rates the utility charges for that electricity. The rub here, however, is that the utilities are in the business of selling electricity and realize that your project will not forever be the lone tree standing in the plain, but soon part of a forest of new infrastructure that will easily support the T & D charges imposed over a greater number of users. Consequently, the utility may lower those expenses for your project just to insure themselves of customers in the future.

Therefore, it may be that a second value proposition is needed to make your project work with DG in spite of utility marketing ploys. That second value proposition could be Combined Heat & Power (CHP). Thomas Edison's first central station generating plant employed combined heat & power when he built a district heating system to use the waste heat from his engine to supply heat and hot water to his neighbors. Of course, the NIMBY process (Not In My Back Yard) precludes central stations from being located in populated centers, so CHP is rarely found in large central station power plants.

But, in a distributed generation system it is practical and possible to incorporate CHP which can provide space heating, domestic hot water, and chilled water for air conditioning to the facility, as well as heating neighborhood or municipal swimming pools. Since transmission losses in delivering electricity amount to 7-10% of the generated power, DG systems avoid those losses. With the conversion of fuel to electricity being in the 35% area, i.e. 9,700 Btus of fuel to deliver one kilowatt hour of electricity (3,415 Btus), the value proposition of CHP raises that efficiency to well over 80%. When the cost of heating and air conditioning are introduced to the value proposition, distributed generation combined with CHP becomes an instant winner in returning the investment to such a system.

The technologies of today lend themselves to such systems. Since environmental concerns will be a factor in implementing

DG, the microturbines, small reciprocating engine driven generators, fuel cells and photovoltaics all meet local air quality management district requirements, albeit with catalytic conversion systems on the reciprocating engines. Medium sized gas turbines also show promise in meeting air quality needs. The microturbine field encompasses proven technology in the 60 to 100 kWh sizes and sizes up to 250 kWh are soon to be marketed. These systems allow for multiple units to be operated under one master control and are ideal for modular clusters in whatever facility or development is contemplated.

While fuel cells and photovoltaics are still not being mass produced, their costs may be prohibitive in returning a reasonable investment in all but housing developments where 20-year paybacks could be tolerated. However, at installed costs of $1200-$1500/kWh for microturbines and packaged reciprocating engine cogeneration systems, the returns on investment are in the 4-year and less payback arena.

The term "spark spread" is often employed to determine if an on site generation system is viable. That's the spread between the cost of the local utility's energy and demand charges vs. the cost of fuel and maintenance to operate an on site system. In California and the North East, that spread is sufficient to make straight DG viable even in areas now served by the local public utility.

The discussion on reliability lends itself to DG implementation for not only stand by power but the combination of full power with standby capability, often referred to as Dual Mode. In those cases the on site generator would be connected to the grid and a transfer switch device would sense a grid failure and bring the generator back on line in a stand-alone mode. There may be a delay in accomplishing this therefore a UPS (Uninterrupted Power Supply) device may be needed in the form of batteries to allow for the momentary down time of the generator. This delay is needed in order to confirm to the utility that the on site generator has been disconnected from the grid. This differs from pure standby generators that are never connected to the grid, but react to a grid outage.

That being the case, interconnection agreements are required by the utility. This has created technical, economic and procedural barriers to a smooth implementation of DG. Utilities want to be assured that the proper protective relays are in place to prevent a supposedly "dead" power line on their system from becoming "alive" when working on the line. Most technology manufacturers have these protections and anti-islanding devices built in to their system electronics. Many have pre-certified their technology with the local utilities to prevent duplication every time a new system is installed.

The cost of the interconnection agreement is to cover the utility's time to study the project and insure the safety and propriety of the system. An on site meter may be required to measure the output of the generator in order to determine the cost of any departing load charges or exit fees—a charge per kWh that seemingly pays the utility for the privilege of not supplying electricity but having enough running reserve to back up the on site generator should it go down. Standby charges are also imposed on the nameplate rating of the generator. These can range from 1.00/kW per month to as high as $8.00/kW per month. So, a 300 kW generator may cost the company as much as $2400 per month to be connected to the grid. Many of these barriers are being mitigated by negotiations as well as legislation in various states.

As late as July, 2003, a new entity was formed in San Diego, CA called Regional Energy Policy Advisory Council. It was formed by disparate members such as Qualcomm, a manufacturer of wireless phone technology; Utility Consumers Action Network, an energy lobby; the City of San Diego and a small business trade association. The purpose of the group is to investigate and develop a regional energy future different from the one envisioned by SDG&E the local de-regulated public utility. The council members have signaled their commitment to break with an energy planning paradigm they say is dominated by a utility with interests that sometimes might not be the same as the public's. With just one dissenting vote the 13-member panel rejected the option of becoming an advisory panel to SDG&E—a role favored by the utility—and instead moved forward with its own regional energy plan.

This Regional Energy Policy Advisory Council would like to see an energy system based on clean, renewable sources of power, many of which could be built in the county. The proposed energy authority would function in much the same way as a municipal utility district. These districts have generally provided lower rates and more reliable service in recent years than such investor owned utilities as SDG&E, PG&E and Southern California Edison. The regional power authority would also have the power to float bonds and construct power projects. It would rival SDG&E and its parent company, Sempra Energy, which also is planning an array of power projects within the region.

According to the article which was written in the local San Diego paper, the Union-Tribune the council has set ambitious energy targets calling for 40% of the region's energy to come from renewable sources such as wind power and sun by 2030. It also wants to reduce per capita consumption of natural gas in an effort to break the dependence on fossil fuels. No mention of cogeneration was made which in itself can be considered "green" energy from that portion of the energy created by the waste heat of combustion. While SDG&E is a member of this new Council, its parent is betting heavily on a scenario for the future that does not involve renewable energy technologies. SDG&E is saying that improvements to the transmission lines can boost the region's electric reliability by allowing the company to tap cheaper electricity generated elsewhere. That elsewhere may be from their new gas-fueled power plants in Mexicali, Mexico; Bakersfield, CA, and western Arizona.

None of the SDG&E plans appears to favor or include Distributed Generation since these new plants would transmit power through the existing "improved" grid. Conversely, none of the Regional Energy Policy Advisory Council plans specifically states an inclusion of DG, but it's likely some DG would be involved from the early reports. The Council must believe that between now and 2030 the cost of the wind and sun powered technologies will compete favorably with more conventional power generating equipment on a cost/kWh basis, otherwise the goals of the Council will be hard pressed to demonstrate reduced prices of electricity to consumers.

It is this author's thinking that with such a controversy brewing between the aforementioned Council and the first de-regulated major utility in the United States that DG will show its mettle to both sides of this equation and become a major player in solving this region's cost and reliability issues.

COMBINED HEAT & POWER

Combined heat & power, cogeneration, is a natural market for the microturbine since it delivers the greatest amount of waste heat compared to other engine technologies. It's modularity and ability to be multi-packed into systems containing up to 200 individual units can provide significant power and heat to serve kilowatt and megawatt sized systems.

Since economics is the driving force behind any project the value proposition of CHP is excellent. Utilizing the waste heat from the turbine the overall thermal efficiency of the microturbine CHP system ranges from 75 to 90%—meaning that for every Btu of fuel input to the turbine 75 to 90% of the fuel input will emanate in useful work. The key factor in any CHP system, therefore, is the ability to utilize the waste heat on site in the form or hot water, chilled water, hot air, or in some cases, steam.

An example of the overall thermal efficiency attainable using microturbines in a hot water producing application is:

- Produce a net of 348 kWh of electricity using six (6) 60 kilowatt units
- Consume 48.24 therms of natural gas to operate the turbines
- Produce 29 therms of hot water energy at 120°F

Thermal efficiency = Kilowatts of electricity + Useful thermal energy/Fuel Input

%   = 348 kWh x 3415 Btu/kWh/100,000 Btu/therm + 29
      therms hot water divided by 48.24 therms of fuel input
%   = 11.88 therms of electric power + 29 therms of hot water
      = 40.88 therms divided by 48.24 therms of fuel input
%   = 40.88/48.24 = 84.74 % thermal efficiency

If this same facility were using utility power which generates
one kWh for every 9700 Btus of thermal input, and produced 29
therms of hot water in a hot water heater which has an efficiency
of 75%, the net thermal efficiency would be:

Fuel Input = 9700 Btu/kWh x 348 kWh
           = 3,375,600 Btus or 33.75 therms

To generate 29 therms of hot water in a heater with 75% efficiency requires 38.67 therms of energy.

So, the combined thermal input to generate 348 kWh of electricity and 29 therms of hot water is 72.42 therms (33.75 therms
equivalent for electricity + 38.67 therms for hot water).

But, we can generate the same amount of energy in a cogeneration plant that consumes 48.24 therms of energy. The difference
in fuel input is 72.42 – 48.24 = 24.18 therms saved with a cogeneration plant per hour of operation. Over an 8760-hour year the
accumulated savings in energy is over 211,000 therms per year.

While hot water is a major component of a CHP system, the
next universal energy need besides electricity is chilled water for
air conditioning. Using absorber chiller technology, where a heat
source causes the evaporation of a liquid under vacuum resulting
in the evaporation/condensation process, heat is removed from
circulating water. This water, chilled to as low as 41°F, is piped to
air handlers to provide building air conditioning. Absorber chillers can use natural gas, hot water, steam, or hot exhaust gases to
create the heat source for the evaporation process.

Microturbines, with an exhaust temperature near 600°F, can
use that exhaust heat to power the absorber chiller. For every 60
kWh electrical output, approximately 21 tons of chilled water at

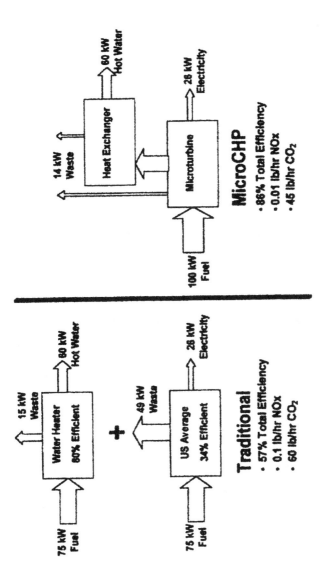

Figure 5-1. Example of Efficiency of CHP vs. Conventional System.

41-44°F can be produced. That eliminates the need for converting the waste exhaust heat to hot water first before sending it to the absorber chiller. Reciprocating engine driven generators produce hot water from the engine coolant and that hot water can be used to power hot water driven absorber chillers.

These same absorber chillers can also produce hot water for hydronic space heating systems by diverting the exhaust gases from the chilled water section to the hot water section thereby utilizing the technology year round in climates that require both air conditioning and space heating.

Using the exhaust gases from microturbines directly into a drying chamber is another example of CHP. The cleanliness of the exhaust gas allows this type of process to be used in applications where large gas turbines with a less clean exhaust cannot be employed without gas clean-up.

BIO-GAS RECOVERY

Bio-gas is the term used to describe methane production from the decay of organic material. Typical bio-gas producers are landfills and anaerobic digesters. In the former case the decomposition of organic matter that has been placed in a mature landfill and covered with dirt starts to form pockets of methane gas that can be recovered by drilling a pattern of holes in the landfill and connecting the holes to a piping system connected to a vacuum system to literally suck the methane out of the landfill. This procedure is often done to prevent free methane from finding its way to the surface and causing air pollution. The resultant collected gas is sent to a flare where the methane is burned. Depending on the efficiency of the flare the destruction may not be complete and the NO_x produced may be a pollutant in itself.

Collecting that gas and putting it through a clean-up system to remove moisture, particulate matter and other contaminants, such as siloxanes, the resultant methane can then be directed to a combustion engine or turbine to produce electrical energy. The Btu

Plastics Factory: Rochester, NY
•*Primary power: 25 30-kW microturbines*
•*Grid is backup*

•*Hot water CCHP:*
•*Process heating*
• *Building heating*
•*HVAC and radiant floor*
• *Heat-driven air conditioning and desiccation*
•*200-ton UTC Carrier chiller*

Figure 5-2. An Array of Microturbines Furnishing Exhaust Heat to a Hot Water Fired Absorber Chiller

© 2003 Capstone Turbine Corporation • www.microturbine.com

Figure 5-3. A Direct Exhaust Fired Absorber Chiller Taking Exhaust Gas From a Microturbine

content of landfill collected gas ranges from 300 Btu/cubic foot to 450 Btu/cubic foot. Carbon dioxide usually makes up the remaining volume of gas possibly mixed with some hydrogen sulfide and water vapor.

Here, again, the microturbine excels. Just like any gas turbine the microturbine can tolerate the inferior quality of the gas with more forgiveness than a reciprocating engine. While hydrogen sulfide up to 7% by volume can be safely burned in a Capstone microturbine, siloxanes should be removed regardless of whether turbines or reciprocating engines are incorporated. Siloxanes are the result of silicone materials decomposing in the landfill. The most common source of those silicones is from the many cosmetics and beauty products that are discarded. Combined with oxygen, the silicone turns into siloxane and when burned as part of the methane can form glass like coatings on the internal parts of the turbine or engine. The recuperator of the turbine is especially susceptible to such a glass coating as is the leading edge of the turbine's power wheel. Carbon filtration is recommended to re-

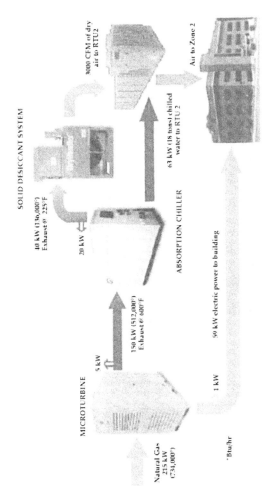

Figure 5-4. Major CHP Components in Producing Chilled Water for Air Conditioning

- *Uncontaminated exhaust stream*
  - *No lubricant vapor because there is no oil or other lubricants*
  - *No oil-fouling of heat exchanger interior or of direct-drying process*
- *Example: Array of 44 C60 (2.6 MWs) at Sanyo Chemical*
  - *Untreated exhaust is ported directly into chemical driers*
  - *Dries polymer product*
  - *Polymer product is used to absorb moisture*
  - *Oil vapor would render the product useless*
  - *Only uncontaminated exhaust heat is usable*
  - *This system has been operating since April '02*

Part of the 44-unit array of C60s at a Sanyo Chemical plant in Japan

Figure 5-5. An Example of Using the Direct Exhaust from Microturbines for Chemical Drying.

move these contaminants.

Refrigerated dryers and gel mesh separators are used to remove the water that is present in landfill gas. Efficient condensate removal systems are imperative in insuring removal of the collected moisture without allowing air to enter the system. Particulate traps are used to collect any solids that may carry over from the collection system. Often, reverse osmosis systems are used to remove the carbon dioxide from the methane if a more pure fuel is desired, but reverse osmosis is an expensive process. There are specialty companies that engineer and manufacture skid mounted clean-up equipment for this service.

Anaerobic digesters also generate methane from the waste product of the "bugs" that literally eat the solid carbonaceous material that comes into a typical waste water treatment plant. The Btu content of the methane in such a digester plant is usually higher than that of landfill methane—approximately 500-600 Btu/ cubic foot. Again, moisture removal is imperative as is any siloxanes found in the system. A gas analysis is a must in determining the quality of the gas and the contaminants it contains so removal can be accomplished to prevent excess maintenance and down time to the microturbine or reciprocating engine.

Since a digester plant operates most efficiently under a higher temperature, the waste heat from the microturbine can be used to generate hot water that is put into the digester to accelerate digestion. Thus, a CHP system is used quite commonly in this type of application. Similarly, the exhaust gas from the heat exchanger can further be utilized as a drying medium to dry sludge from the waste water treatment plant. This could be called a tri-generation system.

Since the fuel is "free" in both landfill and digester gas systems, the value proposition is quite high. In both cases the gas is collected to prevent air pollution and the extra expense to clean it up for combustion in a turbine or engine is a one time capital charge vs. the ongoing purchase of fuel. While digester applications are also CHP applications the return on investment is quite a bit higher than that of landfill applications where cogeneration

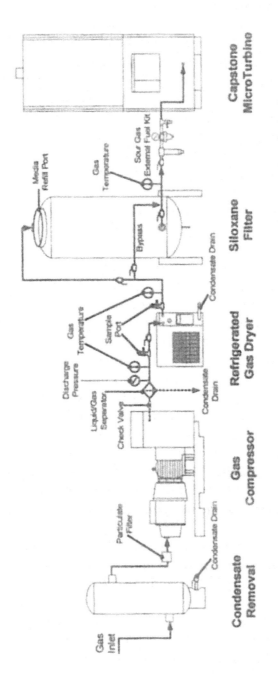

Figure 5-6. Typical Biogas Clean-up Equipment

Figure 5-7. Two Microturbines Exhausting Into An Air/Water Heat Exchanger At a Waste Water Treatment Digester. The Fuel Gas Booster Compressor Is Shown on The Left

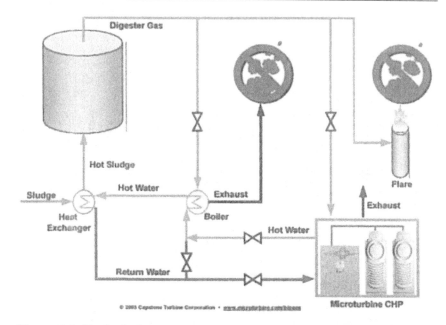

Figure 5-8. Typical Flow Diagram Showing Digester Gas Used To Fuel Microturbines With Waste Heat Recovery Hot Water Going Back to Digester. Note Elimination of Flares

is limited since no hot or chilled water is needed at a landfill. Consequently, the excess power that is generated, after sufficing the on site needs, must be sold to the local utility and often the "avoided cost" price is quite low.

RESOURCE RECOVERY

Another form of "free" fuel is in the oil fields where natural gas is often a by product of oil production. That gas is often flared since no collection system for gathering and storing the gas was put in place. Since most oil field operations require pump jacks or downhole pumps to bring the oil to the surface, electricity is used to operate those pumps. By implementing flare gas burning microturbines to generate electricity, the oil field pumps can run

Figure 5-9. Fifty Microturbines Installed at a California Landfill Burning 350 Btu/Cubic Foot Gas. 1.5 MW of Electricity Produced.

with no grid connected power saving the cost of providing the transmission lines to the remote oil fields. Most flare gas is natural gas at normal heating values and little, if any, clean-up equipment is needed. Since the Capstone microturbine can tolerate up to 7% by volume of H2S, no hydrogen sulfide removal system is required.

One unique feature of the pump jack operation is the regenerative power developed on the down stroke of the pump. The microturbine cannot tolerate this "back feed" of electricity and a capacitor device must be installed to absorb this regenerated electricity when the turbines are operating in a stand-alone mode. If grid connected, the grid absorbs the regenerated electricity.

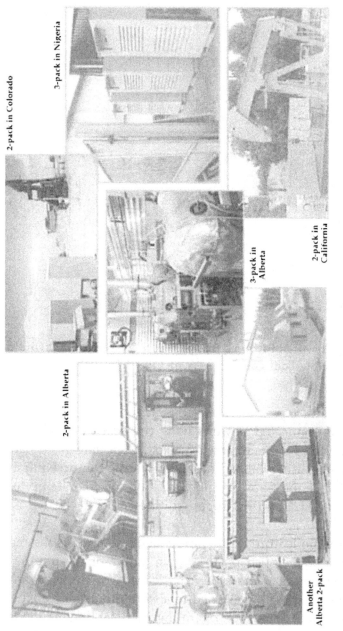

Figure 5-10. Typical Examples of Microturbines in Flare Gas Recovery. Note the Pump Jack in the Lower Right.

- *Remote prime power from a small portion of upstream gas*
- *No air quality permitting issues*
- *Quickly deployed*
- *No fuel trucking*
- *One minor, annual maintenance interval*
- *Reduces O&M cost*
- *Can be easily aggregated to serve loads from less than 30 kW to several MWs*

Operations in Wyoming (top) and in Japan

Figure 5-11. Microturbines Providing Electricity from Coal Bed Methane

In Wyoming, coal bed methane is generated during the mining operation. This methane is of a quality that can be used as fuel in a microturbine to generate electricity that operates the machinery, including the de-watering pumps, associated with the mining operation. The microturbines are generally stand-alone units that provide this electricity without grid interconnection.

PEAK SHAVING

Peak Shaving is the application of on site power production during times of peak power usage, and therefore higher power costs, in order to alleviate that high cost of power. Utilities support this type of application in order to weather the high usage periods when their power grid is operating at its "peak" power production. This usually occurs in the summer time when air conditioning loads are high in warm climates and can also occur in cold climates during peak heating season.

Microturbines lend itself well to Peak Shaving because of their modularity, low emissions, ease of starting and stopping, and no vibration therefore no extensive installation requirements.

The value proposition of peak shaving is usually not as great as that of the other applications mentioned above, but since power reliability is also a consideration during those periods of high electrical usage, the dual mode facet of microturbines is often employed to insure a steady and reliable source of power in the event of a rolling blackout or power outage. This may be considered an insurance policy and therefore the return on investment is incalculable until a power outage occurs. As discussed in the distributed generation section, certain industries require this insurance to protect themselves from severe loss of income.

Chapter 6

Manufacturers of Microturbines

BESIDES CAPSTONE TURBINE CORPORATION there are other significant manufacturers of microturbines producing these mini power
plants. They are:

Bowman Power Systems
Capstone Turbine Corporation
Elliott Energy Systems
Ingersoll-Rand Power Works
Toyota Turbine Systems
Turbec AB

Each will be discussed individually.

BOWMAN POWER SYSTEMS

Bowman offers an 80 kWh unit microturbine totally enclosed
in an acoustically treated package for indoor or outdoor operation.
Bowman focuses on the CHP market and have built into the enclosure the exhaust to water heat exchanger.

The Bowman Model TG80RC-G produces 80 kWh of electricity with a net electrical efficiency of 28% and a potential system
efficiency with CHP of 75% while producing low temperature
water. The unit incorporates a permanent magnet generator,
which Bowman refers to as a turbo alternator, mounted on a common shaft with the compressor wheel and turbine power wheel.

The rotor uses oil sealed bearings and oil lubricated bearings thereby requiring an oil system, with an oil cooler and oil filter. The power electronics are air cooled. The recuperator is external to the main engine with piping taking the compressed air out of the casing into one side of the recuperator while the turbine exhaust is piped to the hot side of the recuperator. High frequency electricity is produced which is brought to 50/60 cycle

Bowman advertised emissions of 25 ppm of NO_x corrected to 15% O_2. Major inspection of internal parts is anticipated at 25,000 hours and to date the high time unit is approximately 10,000 hours. Bowman anticipates 40,000 hours of operation before major overhaul. The package was not UL listed as of February of 2003.

Some of the listed features of the Bowman unit are:

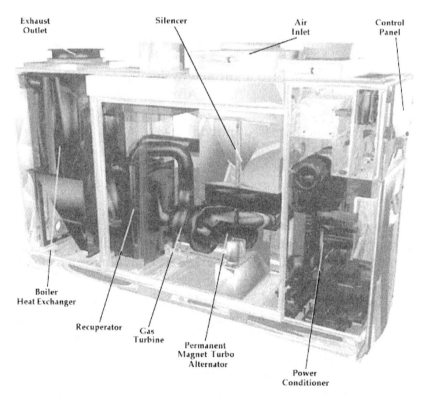

Figure 6-1. Bowman Model TG80RC-G Microturbine

- Fully integrated package
- Stainless steel waste heat boiler
- Single shaft engine rotor
- Utility start
- Excellent power quality via the power conditioner
- Stand-alone capability
- Variety of gaseous fuels i.e. natural gas, LPG, propane and butane
- Quiet operation—70 dba at 1 meter

Bowman Power Systems was founded in 1994 and pioneered in microturbine technology and cogeneration. Their early product was a 45 kWh unit in 1994. They use a Comp Air sliding vane compressor for fuel gas pressure boosting. They feel that the external recuperator is more easily by-passed if higher exhaust temperatures are required. To date approximately 60 Bowman systems have been installed but Bowman anticipates a huge over-

Figure 6-2. The Bowman TG80 CHP Packaged Unit

all microturbine market potential of 40,000 to 60,000 units per year by 2008 mainly in cogeneration, distributed generation, resource recovery and stand-alone units. Recently Bowman Power Systems signed a marketing agreement with Kohler Co., the manufacturer of reciprocating engine generators for stand by service. Approximately 15% of Kohler's 44 distributors have taken on the Bowman line in 2003.

Interestingly, Bowman's engine is supplied by Elliott Power Systems in an OEM relationship. Conversely, Bowman's power electronics is used by Elliott in their packaged microturbines. A 60 kWh unit is planned to be developed and operational sometime in 2003.

CAPSTONE TURBINE CORPORATION

Capstone Microturbines have been used as the focus point of this book since they have the most experience of any of the current manufacturers of microturbines. Founded in 1988 and marketing successful units since 1998, Capstone has over 2500 units in the field and over 4,000,000 hours of combined operation of microturbines.

Presently, Capstone markets the C-30 and C-60 units, nominal 30 and 60 kilowatt units. The published performance of each unit operating on natural gas is shown in Table 6-1.

Both units are exempt in California from the Air Quality Management Districts permitting requirements.

Capstone advertises the main features of their microturbines as follows:

•   Air Bearings requiring no lubrication
•   Air cooled electronics requiring no water cooling
•   Ultra low emissions of NO_x, CO_2, VOC
•   Pure sine wave power output
•   Multi-Packing of units up to 200 units
•   Single rotor design, only one moving part

Table 6-1.

	C-30	C-60
kWh @ Iso rating	30	60
Efficiency (LHV)	26%	28%
Heat Rate (LHV)	13,100 Btu/kWh	12,200 Btu/kWh
Electrical Output	480—3-60	480-3-60
Fuel Usage	433,000 Btu/hr	804,000 Btu/hr
NO_x Production	<9ppmV@15% O_2	<9ppmV@15%O_2
Exhaust Energy	310,000 Btu/hr	541,000 Btu/hr
Noise	65 dba @ 33 feet	70 dba @ 33 feet
Weight	1052 pounds	1671 pounds
Dimensions	28.1"Wx52.9"Dx74.8"H	30"Wx77"Dx83"H
Dual Mode	Yes	Yes
Liquid Fuels	Yes	No
Bio-Gas	Yes	No
UL Rated	Yes	Yes
IEEE 519	Yes	Yes

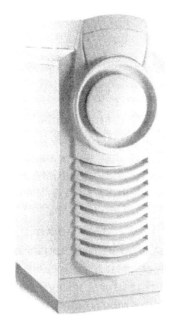

Figure 6-3. Capstone C-30

Figure 6-4. Capstone C-60

- Minimal maintenance requirements
- Utility approved protective relays
- Pre-certification for utility interconnect
- Digital power controller
- Vibration free, quiet operation
- Weatherproof enclosure for outdoor installation
- Rooftop siting
- Built in display and user interface
- Factory installed modem for remote monitoring and operation
- Load following capability
- Multiple gaseous fuel capability by changing fuel index number

Capstone advertises the maintenance requirements on natural gas as air and fuel filter changes every 8,000 hours; injector(s), igniter and thermocouple changes every 16,000 hours; major overhaul in 40,000 hours (minimum). Capstone indicates that based on normal labor and parts costs, the units enjoy a $0.005/kWh (1/2 cent/kWh) maintenance cost.

Capstone's high time unit as of mid-2003 is approximately 39,000 hours on one of the first units placed in commercial operation in late 1998 with no adverse affects being noticeable that a rotor change is imminent. The air bearings have shown no wear on a recent inspection of that unit and the recuperator, which was thought to be the weak link, also shows no need for replacement.

Capstone has announced plans to introduce a 200 kWh microturbine in 2004. A $10 million dollar grant from the DOE in 2002 has helped Capstone fund this project. While no specific details of the larger microturbine have been published, Capstone hopes to maintain the same benefits and excellent robustness in the 200 kWh unit as exemplified in its present line up of microturbines.

INGERSOLL-RAND POWER WORKS

Ingersoll-Rand Company builds their microturbines in a separate division called IR Energy Systems headquartered in Davidson, NC. Ingersoll-Rand shows two models currently being manufactured, the Model 70 SM and the Model 250 SM.

Physical dimensions of the Model 70 are: 71"L x 43" W x 87"H. Weight is 4850 pounds.

The functionality of the IR microturbine has some basic differences from other microturbines due to its split shaft design and more conventional generator. Air is drawn into the inlet of the unit and compressed and then passes through the recuperator that captures the heat from the escaping exhaust gases to pre-heat the compressed air before entering the combustion chamber where it is mixed with fuel. The resultant rapidly expanding gases provide the power by passing through the blades of the turbine wheel to drive the compressor and then through a second turbine wheel called the power turbine. It is the free power turbine, rotating at approximately 44,000 rpm, that drives the conventional two pole generator through a gear reducer.

The rotor is suspended on proven oil lubricated bearings as is the gear reducer.

The hot gases can then enter the optional hot water heat exchanger to capture the waste heat and turn it into hot water in a typical cogeneration application.

The recuperator is of IR design incorporating features of the typical plate and frame heat exchangers but with added design

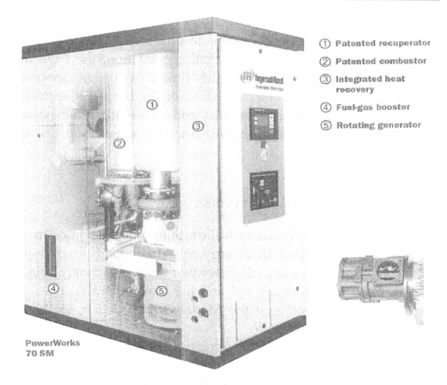

1. Patented recuperator
2. Patented combustor
3. Integrated heat recovery
4. Fuel-gas booster
5. Rotating generator

PowerWorks
70 SM

Figure 6-5. Ingersoll-Rand Model 70 Microturbine (Note the Gear Box Just Above the Generator)

features to overcome the fatigue and durability problems associated with that type of heat exchanger.

Ingersoll-Rand has taken a page from its own book, the KG2 two megawatt gas turbine to develop the Model 250 SM microturbine to deliver 250 kWh of electricity.

Ingersoll-Rand touts its main features and benefits as:

- Industrial grade reliability
- Ten-year system life
- Military grade recuperator
- Integrated fuel gas booster compressor
- Rugged induction generator
- Low emissions < 9ppm NO_x & CO
- Quiet operation

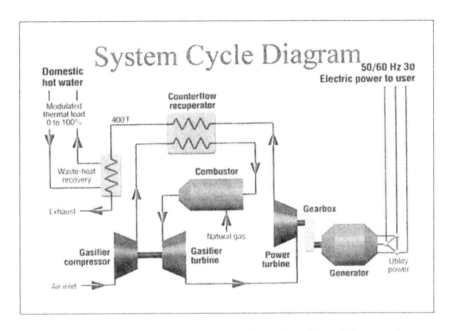

Figure 6-6. The Ingersoll-Rand Microturbine Gas Flow Diagram Showing the Split Shaft Arrangement

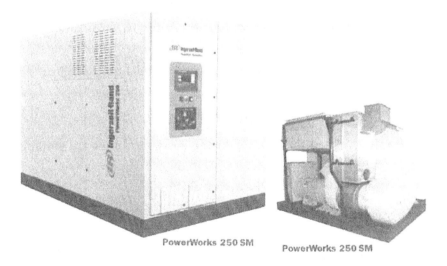

Figure 6-7. The Ingersoll-Rand Model 250SM Shown Both With & Without Enclosure. Physical dimensions of the Model 250 are: 126" L x 76"W x 89.5" H. Weight is 9000 pounds.

Electrical Performance*

Characteristic	Specification
Nominal Electrical Power (±5)	70 kW @ 59°F
Maximum Electrical Power (±5)	92 kW @ 0°F
Voltage	480 VAC
Frequency	60 Hz
Type of Service	3 phase; wye, 4-wire ungrounded
Electrical Efficiency LHV (±2)	28% LHV including fuel-gas booster 29% LHV without fuel-gas booster
Heat Rate HHV	13,550 Btu/kWh including fuel-gas booster 13,080 Btu/kWh without fuel-gas booster

*At ISO Conditions (59°F @ sea level, 60% RH) unless otherwise noted

Figure 6-7a. Performance Characteristics of the IR70 Microturbine

- Fuel versatility with various gaseous and liquid fuels
- Bio-gas operating capability
- Proven oil lubricated bearings

It is advertised that the IR Power Works turbines are suitable only for indoor operation.

ELLIOTT ENERGY SYSTEMS

Elliott is a fully owned subsidiary company of Ebara Corporation of Tokyo, Japan. The Elliott microturbine was developed in Stuart, FL, and is currently assembled there. Elliott offers its Model TA-100 package as shown in Figure 6-8.

Electrical Performance*

Characteristic	Specification
Electrical Efficiency (±2)	32% LHV without fuel-gas booster 31% LHV including fuel-gas booster
Nominal Heat Rate HHV	11,900 Btu/kWh without fuel-gas booster 12,200 Btu/kWh including fuel-gas booster
Nominal Heat Rate LHV	10,700 Btu/kWh without fuel-gas booster 11,000 Btu/kWh including fuel-gas booster
Electrical Power kW (±15)	250 nominal @ 59°F without fuel-gas booster 240 nominal @ 59°F with fuel-gas booster >300 @ 0°F
Voltage	480 VAC
Frequency	60 Hz
Type of Service	3 phase, wye, 4-wire ungrounded
Grid-isolated Regulation (steady state)	±0.50% nominal voltage ±0.3 Hz frequency
Transient Handling (recovery within 3 seconds)	±10% nominal voltage max ±5 Hz frequency max

*At ISO Conditions (59°F @ sea level, 60% RH) unless otherwise noted

Figure 6-7b. Performance Characteristics—IR250 Microturbine

Figure 6-8.

Turbo Alternator™ Specifications

Performance:

Electrical	
Output*	105 kW (+/- 3)
Turndown	100%
Efficiency*	29% (+/- 1) LHV

Fuel Consumption (ISO Rated Power)

CNG Recuperated:	22 SCFM
(@ 1,235,506 Btu/hr.) LHV	

Thermal Output (Hot Water)

	172 kW/587,036 Btu/hr.
Water Inlet Temp	120°F/49°C
Water Outlet Temp	140°F/60°C
Flow	60 GPM
Total System Efficiency*	>75%

Engine Specifications

Manufacturer	Elliott Energy Systems
Model	TA-100
Type	Recuperated Gas Turbine
Pressure Ratio	4 to 1
Fuel Type	Natural Gas

Cooling System

Engine	Oil Cooled
Alternator	Oil Cooled
Inverter	Air Cooled
Enclosure Cooling	3,200 CFM

Exhaust System

Outlet Size	10"
Rated Back Pressure	0 In wc
Max. Back Pressure	5 In wc

Fuel Supply

Pressure Required	0 - 5 PSIG
	0 - 0.345 Bar(G)

Lubrication System

Oil Type	Mobil SHC 824

Oil Capacity with Filters

	5 US gal- (19 L)
Oil Filter	Spin On Type, 3 Micron

Emissions, Natural Gas

CO:	< 41 PPM @ 15% O_2
	< 24 PPM Volume
	54.1 mg/MJ
	52 mg/m^3@ 15% O_2
	1.49 lbs/MWhr
	0.48 grams Bhp
	0.1258 lbs/MMBtu
NOx:	< 24 PPM @ 15% O_2
	< 14 PPM Volume
	51.9 mg/MJ
	50 mg/m^3@ 15% O_2
	1.56 lbs/MWhr
	0.50 grams Bhp
	0.1297 lbs/MMBtu

Exhaust Gas

Temperature	535°F/279°C
Flow	1.9 lbs./Sec.

Batteries

Battery Quantity: (2 Wired in Series)	12VDC min.

* Without Gas Compressor

Figure 6-8a. Elliott Model TA-100 Microturbine Package

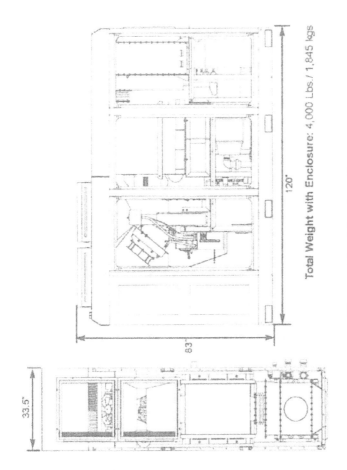

Total Weight with Enclosure: 4,000 Lbs./ 1,845 kgs

Figure 6-8a. (*Continued*)

When Elliott first decided to enter the microturbine manufacturing business in 1993 their first thoughts were for a small gas turbine driving a conventional generator through a reduction gear. The concept, in their opinion, required many moving parts and did not take advantage of the low maintenance costs that microturbines were purported to offer. Consequently, Elliott decided to take advantage of high speed alternator and inverter technologies to increase the electrical efficiency of the turbine and reduce maintenance costs.

Note that in the cross section of the Elliott TA-100 the shaft is supported on oil lubricated bearings requiring oil lubrication with the associated oil supply system.

This configuration is typical of the majority of the microturbine manufacturers i.e. a single moving part with one shaft supporting the generator, compressor wheel and turbine power wheel. Elliott's speed is 68,000 rpm vs. Capstone's 96,000 rpm. As mentioned in the treatise on Bowman Power Systems, the Elliott engine is used by Bowman in their unit with Bowman's power electronics used in the Elliott package.

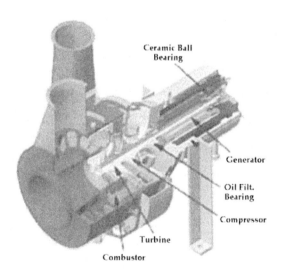

Figure 6-9. Cutaway of the Elliott TA-100 Microturbine

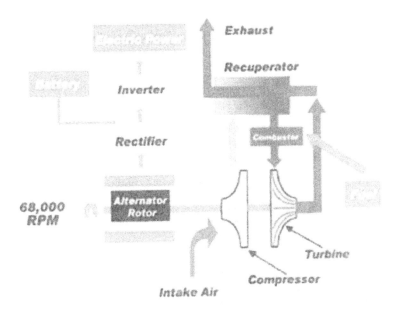

Figure 6-10. Schematic of the Elliott TA-100

Elliott has had many false starts in the marketing of their product. At one time General Electric had been a partner with Elliott, but due to the inability to bring the unit to market, that partnership was dissolved and Ebara, who had a small stake in Elliott, became the full owner of EESI.

TURBEC AB

Turbec, a consortium between Volvo and ABB, builds a 100 kWh, packaged microturbine with built in hot water heat exchanger for the CHP market.

The development of the Turbec unit was begun within Volvo in the late 1980s when a 40 kW prototype was built. This unit was demonstrated in the hybrid automobile in 1992 with a microturbine-battery system. Within Volvo a 100 kW unit was built for the truck and bus markets.

Volvo realized that the hybrid vehicle market was somewhat

Figure 6-11. The Turbec T100 CHP Packaged Microturbine

limited while power generation showed virtually no limits. Since the power electronics to bring such a unit into that market was needed, the partner that Volvo sought was ABB, who could provide that technology. That's when the partnership moved from Volvo to Turbec in 1998.

The Turbec unit shows similarities to the Capstone and Elliott/Bowman systems in that it has a single rotor design with built in recuperator. However, the package also incorporates the air to water heat exchanger for hot water production, something Capstone only developed in 2003.

Turbec considers its more important features and benefits in the way in which they integrate the recuperator to increase the electrical efficiency; their combustion technology leading to very low emissions for gaseous as well as liquid fuels; and their overall robustness of the engine. To date, Turbec does not have a unit capable of operating without grid presence, i.e. stand-alone, since the European market has not called for such features. Also, due to the overall weather conditions in Northern Europe, the T100 is not suitable for outdoor application.

While Turbec's advertised NO_x emissions of 15 ppm is a bit higher than that quoted and guaranteed by Capstone and

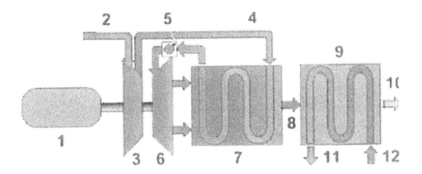

1.	Generator	7.	Recuperator
2.	Air inlet	8.	Exhaust gases
3.	Compressor	9.	Heat exchanger
4.	Air to recuperator	10.	Exhaust gas outlet
5.	Combustion chamber	11.	Hot water outlet
6.	Turbine	12.	Water inlet

Figure 6-12. The Turbec T100 CHP Schematic

Ingersoll-Rand, they are working on catalytic combustors to achieve levels as low as 3 ppm with a higher turbine exhaust temperature into the catalytic combustor of 1350°C using ceramic technology. Today's metallic microturbine is typically around 1000°C. The Volvo influence in automotive catalytic combustors is evident here.

Turbec believes that the single shaft engine with high speed generator has completely different characteristics compared to a two shaft engine. The advantages for the high speed generator are cost, reliability, and quality of electricity.

While Turbec is a European company it desires to enjoy a key marketing presence in the United States. It envisions the CHP and DG markets as key to becoming one of the leading microturbine companies in the world. To accomplish this goal requires a rather large market share. However, some of the barriers to this goal are some of the utility certifications required in the U.S. for interconnection such as IEEE519 and UL underwriting, neither of which Turbec has as of mid 2003.

Table 6-2. Turbec T100 CHP Performance Specification

General Identification

Usage:	Indoors
Dimensions of CHP unit:	Width 870 mm (33.9")
	Height 1,900 mm (74.8")
	Length 2,920 mm (115")
Weight:	2,000 kg (4470 lb)

Gas turbine

Compressor type:	Centrifugal
Turbine type:	Radial
Type of combustor:	Lean pre-mix, low emission
Number of combustors;	1
Pressure in combustor:	4.5 bar (A) 65 (psia)
Number of shaft;	1
Nominal speed:	70,000 rpm
Consumption of lubrication oil:	<9 litre/6,000 h operation
	(<304.9 ft oz 76,000 h op.)

Electrical data

Voltage output:	400 VAC alt. 480 VAC, 3 phase
Frequency output:	50 Hz alt. 60 Hz

Fuel requirements

Pressure min/max*:	6/8.5 bar (g) 87/123 psia
Temperature min/max*:	0°C/60°C (32°F/140°F)
Lower heating value:	38-50 MJ/kg
	*without fuel gas compressor

Fuel gas compressor

Gas suction pressure:	0.02 - 1.0 bar (g) (0.3 - 14.5 psi)
Compressor type:	Scroll compressor
AC power supply	345-525 VAC, (50/60 Hz)
Noise level	75 dBA at 1 m (3.3 ft)
Dimensions:	Width 610 mm (24")
	Height 1,070 mm (42")
	Length 1,370 mm (54")

Performance data

Net electrical output*:	105 kW (±3)
Net electrical efficiency*:	30% (±1)
Net total efficiency*:	78% (±1)
Fuel consumption:	350 kW (1,194,000 Btu/h)

(Continued)

Table 6-2 (*Continued*). Turbec T100 CHP Performance Specification

Net thermal output (hot water):	167 kW (±5)(570,000 Btu/h)
Exhaust gas flow:	0.80 kg/s (6,350 lb/h)
Exhaust gas temperature:	85°C (185°F)
Water inlet temperature:	50°C (122°F)
Water outlet temperature:	70°C (158°F)
Noise level:	70 sBA at 1 meter (3.3 ft)
Volumetric exhaust gas emissions at 15% O_2	100% load
No_x	<15 ppm/v = 32 mg/MJ fuel
CO:	<15 ppm/v = 18 mg/MJ fuel

All performance data at ISO conditions and at 100% rated load
(Fuel gas compressor power consumption excluded)

Maintenance
This simple and rugged design of the T100 power module provides for a durable operation during many years. Expected lifetimes of main components are listed below:

Gas turbine engine:	> 60,000 hrs
Recuperator:	> 60,000 hrs
Combustor:	> 30,000 hrs
	(some parts < 30,000 hrs)

The preventive (scheduled) maintenance is divided into two different categories:

	Interval (h)	Outage (h)
Inspection	6,000	24
Overhaul	30,000	48

TOYOTA TURBINE AND SYSTEMS, INC.

Toyota is a relative newcomer to the microturbine business introducing their 50 and 350 kilowatt units as pictured below.

The Toyota microturbine is not yet marketed in the United States, or in any other foreign country, as of August 2003 and no plans have been announced as to when its overseas introduction may be made.

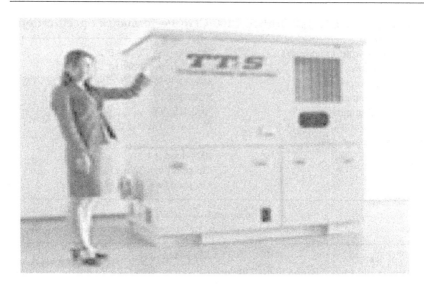

Figure 6-13. Toyota Turbine Systems Model TTS-50 Cogeneration Package

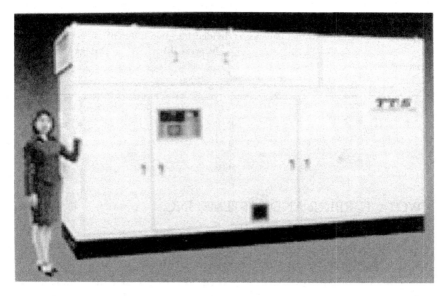

Figure 6-14. Toyota Turbine Systems Model TTS-300 Cogeneration Package

Chapter 7

Operating Histories

THIS CHAPTER DISCUSSES some of the operating histories in various applications of microturbine technology. It is not meant to highlight any one manufacturer but to show the different venues in which microturbines can be deployed.

HARBEC PLASTICS FACTORY—
NEW YORK, USA

This company had been hit by frequent power problems and wanted a more reliable power supply as well as lower energy costs. Further, the owner also wanted a more environmentally friendly impact compared to traditional utility power and on-site gas heating. The solution was the microturbine installation with a high level of redundancy. Harbec installed twenty-five (25) Capstone 30 kilowatt microturbines with a total nominal 750 kW max electric output and up to 52 therms of hot water output. This project was commissioned in 2000.

To achieve high energy efficiency, 20 of the microturbines are connected to air to water heat exchangers to produce hot water and some steam. Part of the hot water is used during the summer to supply an absorbtion chiller.

The units are dual mode, thereby able to produce electric and thermal energy in the event of a utility power outage. Even though grid connected, the units are Harbec's primary source of electrical energy.

Figure 7-1. Harbec Plastics Factory Array of a Microturbine Cogeneration System

BRICKWORKS, UNITED KINGDOM

This company has a need for direct heat in their brick-drying kiln. By installing two Bowman 80 kilowatt TG80SC, simple cycle microturbines fueled by natural gas, they were able to match the desired kiln temperature by using the exhaust from the micro-turbines directly into the kiln. Being simple cycle, i.e. non-recuperated, the exhaust temperature is over 1000°F. Almost twice as much natural gas is used to fuel the microturbines, but is offset by the natural gas the kiln required while producing 160 kilowatts of electricity.

The velocity of the exhaust gases maintain good thermal pattern configuration within the kiln thereby increasing the throughput of bricks while saving energy. The kiln exhaust gases are recycled, i.e. returned to the process as warm input air. The

Figure 7-2. Two (2) 80 Kilowatt Microturbines Supplying Direct Exhaust Heat

total thermal efficiency of the system is over 90% and the resultant payback was about 2 years. The system was commissioned in September of 2001.

GREENHOUSE, SWEDEN

Four million cucumbers per year are the beneficiary of a 100 kilowatt cogeneration installation at a greenhouse in Sweden. Powered by a Turbec T-100 unit with an integrated hot water heat exchanger plus an additional heat exchanger to reduce the exhaust gas temperature before spreading it in the greenhouse, the overall thermal efficiency is above 90%. Thanks to the heat and the CO_2 from the microturbine the cucumbers grow 30% faster.

Despite relatively low electric prices the combination of increased cucumber production due to the hot water and CO_2, as well as offsetting relatively high natural gas prices, the system enjoys a 4.2-year payback. It was commissioned in August of 2001.

Figure 7-3. A One Hundred Kilowatt Microturbine Cogeneration System Operating in a Greenhouse in Sweden

BIOGAS AT LANDFILLS AND WASTEWATER TREATMENT PLANTS

As discussed in previous chapters, biogas is the name given to methane production emanating from anaerobic processes from biomass, wastewater, landfills, livestock manure and agricultural waste disposal. Since these sources are renewable, power produced from biogas is considered a renewable energy.

A landfill in Jamacha, CA, utilizes four (4) Ingersoll-Rand Model 70S microturbines each producing 70 kilowatts while burning gas with as low as a 30% methane content. Less than 10% of the electric power produced by the project is used to satisfy on-site requirements of the landfill. The remainder of the power is sold to San Diego Gas & Electric Company under a long term power sales agreement. The project was aided by securing a California Energy Commission grant to assist in financing.

**Figure 7-4. Microturbines Operating
at a Landfill in Jamacha, CA.**

Besides the four microturbines the project incorporates a 200 scfm blower/refrigeration skid which pressurizes the landfill gas and reduces moisture and other impurities in the raw gas being collected. Switchgear and a 480v/13.8kv step up transformer and a plant control computer which allows remote monitoring and operation also make up the installation. The project was commissioned into commercial operation in January 2002. No payback period was advertised.

Conversely to the landfill application, methane from wastewater treatment plants which use the anaerobic method of digestion, produce a higher Btu content of waste gas, use all the electricity produced by on-site generators and cogeneration is evident by the need for hot water inside the digester to accelerate digestion.

In addition to the economic benefits, the environmental benefits of capturing the methane are very significant. Methane is a greenhouse gas over 20 times more negative for global warming that carbon dioxide.

Figure 7-5. Five Microturbines at a
Waste Water Treatment Plant in Japan

OIL AND GAS

Many oil and gas exploration and production sites represent
significant opportunities for resource recovery by capturing the
waste gas emanating from such oil operations that usually are
burned in a flare. Whether vented or flared, the gas is not only a
waste of energy but also an environmental problem. While large
gas turbines and reciprocating engines are used at multi megawatt
site, microturbines are extremely suitable for smaller and dis-
persed sites.

This application is very demanding as the sites are frequently
remote, weather conditions vary and the gas is not conditioned.
Containerized enclosures are common. The low emissions from
microturbines may add to the value proposition due to local environmental restrictions or penalties for venting or flaring this
waste gas.

Well head gas from offshore drilling platforms can be burned
in small microturbines to produce power aboard the platform.
With the microturbine so tolerant of H2S and other contaminants,

Figure 7-6. Skid Mounted Two-Pack of 30 kW Units in Alberta, Canada

only moisture removal is needed to clean up the gas. Using the on-site fuel is a major advantage at remote sites since the alternatives of supplying diesel fuel to the site can be very cumbersome and expensive.

In the picture shown in Figure 7-7, the C-60 is near the edge of the platform in the rear on the left, while the C-30 is the smaller unit on the right. Redundancy of units is advisable to insure no interruption of power supply.

PEAK SHAVING, STANDBY POWER

Microturbines are not necessarily suitable for emergency standby power since start up of < 10 seconds is required in most instances. This market is dominated by diesel fueled reciprocating engines whose only requirement is to come on line during a power outage. Y2K was an example of depending on this type of power in the event the turning of the calendar to year 2000 created

Figure 7-7. A Capstone C-60 and a C-30 on an Oil Rig in the Mexican Gulf

major upsets in grid power. Of course, that did not happen and many of those standby engines are still in inventory somewhere.

However, microturbines can be successfully used when both standby power and peak shaving is the goal. Peak shaving is when the utility supplied power is very expensive at the "peak" of power usage. This is usually during the afternoons during the summer and early evening and morning during the winter. Utilities recognize the high demand by charging more for power during those times. Therefore, on-site generators can have a significant value proposition by operating during those "peak" periods. However, since the microturbine can also operate during a grid outage, it can act as a source of standby power as well. Diesel units are limited to a certain number of hours of operation per year, usually 200 hours, due to the pollution they spew during operation. Although the "switching" time for a microturbine is usually 2 minutes, a small UPS system can get the client through those two minutes for the critical loads, i.e. computers and emergency lighting. Rather than purchase a 200 kW diesel generator for $125,000, purchase three (3) 60 kilowatt microturbines complete with standby features for about $175,000 and have continuous power available when the value proposition allows.

POWER QUALITY & RELIABILITY

A lesser known application for microturbines is Power Quality and Reliability. VAR support is one of the more common solutions incorporating the need for better power quality. VAR = reactive power. VAR support is a popular abbreviation for reactive power compensation. The main purpose of VAR support is to reduce electric losses and to increase power line capacity.

Reliability can be enhanced by using multiple units as well as by combining the microturbines with grid power. For example, in

Figure 7-8. A 4-Pack of Microturbines Installed at an Electric Distribution Substation

cases of a weak electric grid with frequent interruptions, customers may choose to have microturbines be the primary source of power for prioritized loads and let the grid be the backup.

As mentioned in Peak Shaving, the ability of the microturbine to operate as both full power and standby power, the use of these units with UPS systems can provide seamless transition to on-site power in case of a grid outage.

Power quality is also a feature of the microturbine with power electronics as its main power producing feature. The pure sine wave produced by such technology is important to sensitive power users in this age of technological gadgets.

HYBRID ELECTRIC VEHICLES

This is where the microturbine was thought to be the main source of application. Toyota, Turbec and Capstone to a large degree started their respective microturbine development for automotive applications. High power density, ultra-low emissions without any catalytic converters and the potential for low maintenance were factors contributing to the interest in microturbines for automotive use.

Presently, only Capstone is active with microturbines for automotive use. More precisely, they target heavy-duty hybrid electric vehicles for urban traffic, e.g. buses and other fleet vehicles. Their ability to operate even on diesel fuel while still meeting the most stringent emission requirements, compact design, low maintenance and long life are attractive attributes of the microturbine for this market.

The system configuration is series hybrid, i.e. the electric power for the propulsion of the vehicle comes from an energy source which in almost all cases is a battery. The microturbine is the on-board battery charger and provides range extension. In many cases the system is optimized in such a way that the vehicle can operate without interruption during day time and then recharged during night time using grid power.

Figure 7-9. Example of Hybrid Electric Vehicle Use of Microturbines

The lack of oil lubrication in the Capstone microturbine, thanks to air bearings, provides a simpler system with less maintenance to make this application a decided winner.

HYBRID SYSTEMS

The phrase, "hybrid systems" is a popular but not completely accurate description of combined systems with a microturbine and another power generating device. The purpose, of course, is to achieve a significantly increased electric efficiency. This application is at an early stage but examples exist where two supposedly different energy producing technologies are operating together to accomplish the goal. The National Fuels Research Center in Irvine, CA, is demonstrating a system using a Siemens-Westinghouse 100 kilowatt solid oxide fuel cell (SOFC) and an Ingersoll-Rand microturbine. The fuel cell gives off waste heat that the microturbine uses to generate another 100 kilowatts of electricity. No internal combustion occurs within the microturbine.

34 feet HEV bus for propane by Desighline, New Zealand.

Right: 15 ton HEV locomotive for diesel fuel by Tomoe, Japan

Below: 38 feet HEV bus for diesel fuel by Chargeking, China

Figure 7-10. Examples of Hybrid Electric Vehicle Usage by Microturbines

Fuel Cell Energy has demonstrated as system using their molten carbon fuel cell (MCFC) and a Capstone simple cycle microturbine.

These combinations of fuel cells and microturbines provide the most common aspect of "hybrid system." They rival the combined cycle gas turbines for efficiency, exceeding the 55% electric efficiency of those plants. With no "internal combustion" the cleanliness of these hybrids is ultra clean meeting any Best Available Control Technology in existence. Of course, the high cost of the fuel cell is still the major drawback to hybrid systems. The 200 kilowatt system at the National Fuels Research Center cost over 16 million dollars or over $80,000 per kilowatt, but it was and still is an experimental system.

Other hybrid configurations are possible with the emanation of heat as the necessary by-product in order for hybrids to work.

Figure 7-11. Molten Carbon Fuel Cell Operating with a Capstone Microturbine

INDUSTRY DEVELOPMENT TRENDS

While microturbines have only been successfully marketed since 1998 and are considered a young technology, the fact that individual companies have experienced over 4,000,000 combined hours of operation and close to 40,000 hours of operation on several systems, the robustness of the microturbine has been proven.

The companies that came after Capstone showed the need for integral CHP systems was marketable and Capstone followed suit. Larger engines are needed and the industry is responding thanks to DOE grants to Capstone, Ingersoll-Rand and GE (a company that is not included in this book since they are not yet in production) to produce a 200 kilowatt unit in Capstone's case and the already marketed 250 kilowatt unit in I-R's case.

Increased efficiency is also being sought with Turbec leading the pack with their T-100 exceeding 30% electric efficiency, but the Advanced Microturbine Program funded by the DOE in 2001 is targeting 40% electric efficiency. If and when that is achieved it will rival the electric efficiency of the reciprocating engine.

While microturbines now burn most fuels that are commonly found, including the tough low Btu fuels found in the biogas field, other fuels are being investigated such as hydrogen and ethanol.

Lower emissions are being sought in order to comply with even more demanding future standards. California's 2007 emission standard is very close to a "zero emission" level. It will most likely require catalytic combustion techniques although in actual tests of the current microturbines, less than 3 ppm of NO_x is being demonstrated, even though guarantees of less than 9 ppm are offered. Capstone recently decided not to certify their C-30 to meet California Air Resources Board (CARB) standards, yet the C-30 was the very unit that the Air Quality Districts in California exempted from regulation due to its clean burning aspects.

As these goals are attained and the microturbine market increases, as predicted by Bowman and Turbec, the cost per kilowatt of these Mighty Minis will come down and the proliferation of microturbines will meet early predictions in all the applications discussed here.

Chapter 8

*Realism in the Future**

THIS CHAPTER DISCUSSES the use of microturbines to provide a low cost, high efficiency, precise power system. It should be extremely reliable even though it eliminates the need for a redundant generator set.

Many devices such as computers cannot tolerate even short interruptions in the supply of electricity or even slight deviations from rigid quality specifications for power. Many manufacturing processes such as semiconductor fabrication have similar requirements. Power interruptions or poor quality power can result in loss of data, computers going off-line, damage to hardware, loss of in-process product, delayed product deliveries and other problems.

The most common solution to providing continuous high quality power is to use a static uninterruptible power system. Generally, this system rectifies utility power to direct current and then inverts it back to the required frequency, either 50 Hz. or 60 Hz. Thus the critical loads are isolated from any transient spikes or problems on the utility buss. A stored energy source such as batteries float on the dc buss allowing the system to ride through interruptions for typically five to fifteen minutes. This permits the critical loads to be shut down in an orderly manner or gives time for standby generator sets to be started and brought on line"

An alternate solution is to dispense with the static system and use generator sets to provide precise power to the critical loads and only to the critical loads. Thus the critical buss is isolated from utility problems such as switching transients and light-

*A paper, "Precise Power Systems," by Robin Mackay, September 2002.

ning strikes as well as from in-house problems such as power surges or voltage dips caused by motor starting inrush. Generally this has meant the installation of an "n+1" system. In other words, one more generator set than is necessary to meet the maximum load is installed. Thus, if a generator set fails, the remaining generator sets can still handle the load.

A major deficiency in "n+1" systems is the increase in capital cost due to the redundant generator set. Thus if two generator sets can handle the load, three must be installed for roughly a 50% cost increase. A worse case occurs if one generator set can handle the load. Then the addition of a redundant generator set essentially doubles the cost of the hardware, the installation and the floor space.

A second deficiency is that the generator sets do not operate at their full rating as they must always be prepared to increase power to handle the required load when a generator set fails. Indeed, if one generator set can handle the load and two are installed, the two generator sets must operate at no more than half of their rating if they are to be able to pick up the full load when a generator set fails. With conventional single shaft, gas turbine-driven generator sets, this has meant a dramatic decrease in fuel efficiency. Correspondingly, microturbines must either operate at much higher speeds than is optimum for efficiency if they are going to be able to assume the increased load when a generator set fails or they must use stored energy such as batteries to assist in picking up the increased load.

However, the proper use of microturbines removes the need for a redundant generator set and that is the subject of this paper. The vast majority of microturbines in the field today consist of modest sized gas turbines that drive generators that operate at the same speed as the gas turbine. These very high-speed generators operate at dose to 100,000 rpm to produce power at very high frequencies in the 1600 Hz range. As this frequency is too high for most applications, a power conditioning system is needed. The generator output is rectified to direct current and then inverted to conventional frequencies such as 50 Hz. or 60 Hz.

The concept presented here proposes to have two inverters fed by the rectifying power supply. The first or critical inverter would feed the critical loads and the second or non-critical inverter would assist the electric utility in feeding the noncritical loads.

It is common practice to use bi-directional inverters for both the rectifying power supply and the output inverter. In applications where the output is to be paralleled with the utility grid, this permits the power conditioning system to operate in reverse when the gas turbine is to be started. Thus utility power can be rectified to direct current by what would normally be the output inverter. Correspondingly, what would normally be the rectifying power supply can then invert the direct current and provide alternating current to the gas turbine's generator that now becomes the starter motor.

Microturbines like conventional generator sets can usually operate in two separate modes. They can parallel with the utility grid and deliver a fixed amount of electricity to the utility buss. They can also operate in a stand-alone mode in which case the amount of electricity that they deliver matches the load that they are supplying and vary as the load changes.

Figure 8-1 shows the equipment and the wiring connections. In all figures, the control wires are shown in light dotted lines. In figure 8-1, the power wires are shown in heavy solid lines. In the remaining figures, only the wires transmitting power are shown and they are shown in heavy solid lines with arrowheads to indicate the direction of power flow.

Figure 8-2 shows the normal operation of the system. The gas turbine fuel control would be set in grid-parallel mode and provide a fixed amount of fuel and therefore a fixed amount of power from the generator. This is the most common mode of operation for any generator set operating in parallel with the utility grid. Generally the setting would match the maximum continuous power specified by the manufacturer for specific conditions such as altitude and inlet temperature noting that changes in conditions might change this fuel setting. Most gas turbines would therefore

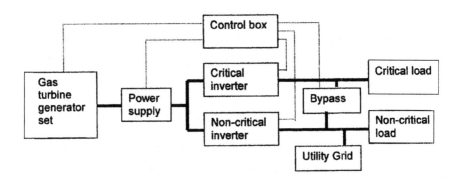

Figure 8-1. Wiring connections

operate at their most efficient power setting and achieve the maximum possible load factor.

The critical inverter would operate in stand-alone mode providing high quality power to the critical loads. The output of the inverter would follow the load as the critical loads change. The output would be isolated from electrical problems on the non-critical buss. The non-critical inverter would operate in grid-parallel mode providing power to the non-critical loads in conjunction with power from the electric utility. This inverter power would reduce the amount of power being purchased from the electric utility or in rare cases feed power back into the grid.

As the power output of the gas turbine's generator is fixed, the total power produced by the two inverters would also be fixed. Thus if the critical load increased, the power delivered to the non-critical buss by the non-critical inverter would decrease and vice versa. This would happen automatically using signals from current transformers on the generator output and on both of the inverter outputs. The output of the non-critical inverter would be adjusted to match the output of the generator set minus the output of the critical inverter with appropriate corrections for efficiencies and any parasitic losses. This is a similar to the control techniques used by any microturbine paralleled with the utility.

Figure 8-3 shows the operation of the system during a utility power failure. If the utility power failed, the non-critical inverter would disengage from supplying power to the non-critical loads.

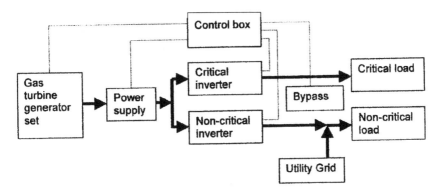

Figure 8-2. Normal Operation

This would reduce the load on the gas turbine. The gas turbine fuel control would change from fixed power or grid-parallel mode to stand-alone mode and act as a conventional generator set does when operating in stand-alone mode and not paralleled with the utility grid. The only loads on the system would be the critical loads, as the non-critical inverter would no longer be providing power to the non-critical buss. The gas turbine generator would then load-follow the critical loads and provide high quality power. When utility power was restored, the non-critical inverter would parallel with the utility grid using conventional techniques. The gas turbine fuel control would revert to fixed power and normal operation would resume.

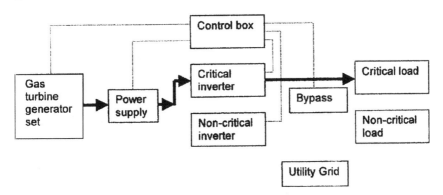

Figure 8-3. Utility Grid Failure

Figure 8-4 shows the operation of the system if the gas turbine were to fail or if the fuel supply were to fail. The gas turbine would stop and the generator would cease to produce power. Then the bi-directional non-critical inverter would reverse its function. Instead of taking power from the power supply and delivering alternating current to the non-critical buss, it would take power from the electricity utility and power the critical inverter. Thus the critical inverter would operate in stand-alone mode and continue to provide quality power to the critical loads without interruption. The critical loads would still be isolated from utility or in house electrical problems. As soon as the gas turbine could be restarted, normal operation would resume.

Figure 8-5 shows the operation of the system during restarting of the gas turbine. Power from the utility grid is rectified to direct current in the non-critical inverter and then inverted back to variable frequency, variable voltage current in the power supply. This power is used to start the gas turbine with the generator acting as a starter motor. This is common practice for starting microturbines where utility power is available. If utility power is not available, batteries can provide power to the direct current buss and the power supply can provide the appropriate power to start the gas turbine. This is common practice for starting microturbines where utility power is not available.

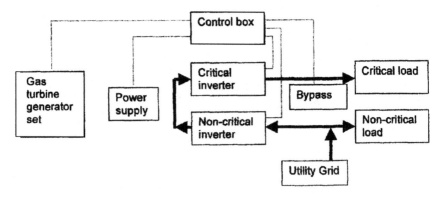

Figure 8-4. Gas Turbine Failure

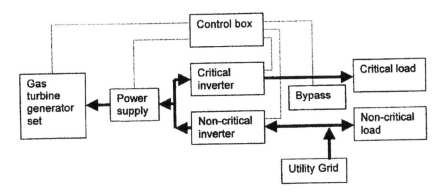

Figure 8-5. Restarting the Gas Turbine

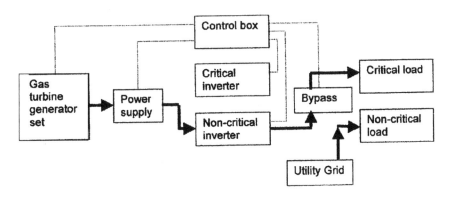

Figure 8-6. Critical Inverter Failure

Figure 8-6 shows the operation of the system if the critical inverter fails. The critical inverter operates in phantom parallel with the grid so that the critical load normally has the same voltage, frequency and phase angle as the utility grid. Thus the critical load can be transferred to the grid through the use of a static bypass switch. However, the non-critical inverter is still operational and being fed by the generator set through the power supply. Therefore, the non-critical inverter would feed the critical load through the bypass switch. The output of the non-critical inverter would then be isolated from the grid which would continue to feed the non-critical load.

Figure 8-7 shows the system being bypassed as might be necessary during system maintenance. Note that during normal operation, the critical inverter would operate in phantom parallel with the utility grid. In other words, the frequency and phase of the output of the critical inverter would be identical to that of the utility grid. Thus, in the event of a complete system failure, the critical loads could be transferred to the utility grid without interruption. This transfer could be facilitated by the use of static switches in the bypass.

This system has several advantages over the "n+1" system described above. It is less expensive to buy because one gas turbine is eliminated. It is more fuel-efficient as the gas turbine operates at higher loads. The utilization is higher because of the additional power delivered by the non-critical inverter. The power delivered to the non-critical buss reduces the electric utility bill. Finally, the system offers additional reliability in that it can operate from either of two energy sources—fuel from pipelines or storage, or electricity from the utility.

The above description assumes that the system is based on a single microturbine powering a single power supply which in turn powers two inverters, one for the critical load and one for the non-critical buss. However, it should be noted that there are numerous microturbine installations utilizing multiple microturbines in which the electrical output from their inverters is paralleled.

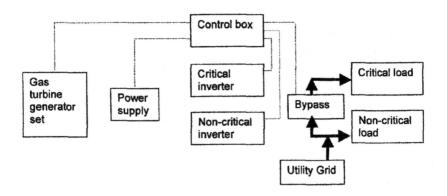

Figure 8-7. System Isolation

Correspondingly, it is feasible use multiple microturbines and parallel the critical inverters to feed the critical loads while paralleling the non-critical inverters to feed the non-critical buss.

Figure 8-8 shows a three-unit installation but essentially any

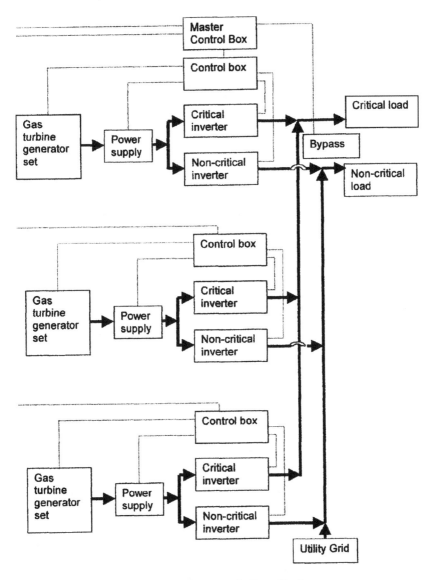

Figure 8.8 Three-unit Installation

number of units could be paralleled. A single bypass switch is shown which could connect the utility to the critical loads in the event of a system failure or if system maintenance is required. It is also feasible to have individual bypass switches for each unit, which would generally act simultaneously.

The precise power system as described above is patent pending and is used here with permission from Mr. Mackay. It is his desire to see this innovation incorporated into a commercially operating microturbine system in the near future.

ALM TURBINE, INC.

This book would not be complete without an introduction to the unique microturbine design of ALM Turbine, Inc. that is building a prototype unit in 2003, but has not yet produced a commercial version.

They are developing a new technological approach to increasing efficiency and performance over conventional microturbines by utilizing Exhaust Gas Recirculation (EGR), with counter rotating turbines and a novel blade cooling technology. They envision efficiencies of up to 40%. Another innovation is to control the power by varying the density of the combustion gases via recirculation. This approach makes the ALM engine highly scalable and provides better responsiveness to sudden load and speed changes. Stable efficiency over a broad range of loads is accomplished making this design suitable for vehicular use.

The gas turbine engine EGR recirculation ratio is about 6 to 1, meaning that only 1/6 of the total gas flow is introduced as new flow. The remainder is already combusted gases. With less air flow being added to the flow duct, less NO_x is produced. ALM feels that the NO_x level will be < 5 ppm. Other goals are to produce from 50 to 300 kWh from the same unit, again by varying the density of the gases via recirculation. ALM hopes to introduce their product in the $600 - $700/kWh sales price.

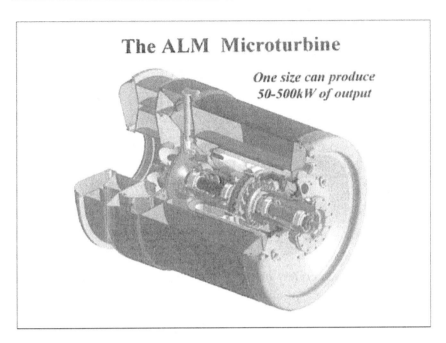

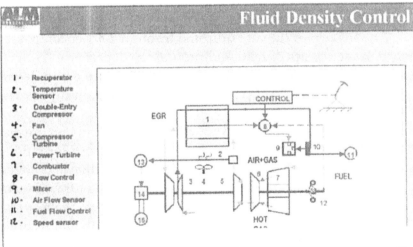

Figure 8-9. The ALM Turbine Cutaway and the Fluid Density Control System

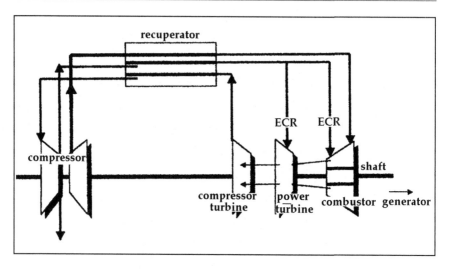

Figure 8-10. An Overview of the Recirculation Cycle in an ALM Turbine

CONCLUSIONS

The history of the microturbine through 2003 encompasses only a scant 5 years of field and production experience. With all manufacturers contributions less than 3000 units are presently operating around the world. Yet, with one major manufacturer ceasing production in 2002, Honeywell Corporation, new entries have come into the market such at Turbec, Ingersoll-Rand and Toyota. Honeywell has not specifically advertised the reasons for pulling out of the market. They had over 300 units in operation and pulled all of them out of the field, reimbursing the clients for their expenditures, and literally destroying the units. Inability to meet low NO_x levels and unable to show maintenance intervals that were acceptable seem to be the main reasons for Honeywell's decision. The irony that Honeywell acquired Allied-Signal companies, the very birthplace of the microturbine via Robin Mackay's and Jim Noe's alma mater, Garrett Corporation, is of curious interest.

Some luck is involved in how and when a new product is brought to market. Capstone nearly pulled the plug early in the game when a unit at Ontario Hydro had a fuel valve fail that inundated the turbine with natural gas when it was shut down. The hot metal in the microturbine started a fire inside the turbine which could have blown with disastrous results. The power electronics controller failed and the fuel proportioning valve had an open signal when sensing a shut down causing fuel to enter the unit when the unit was actually in a shut down mode. Capstone corrected that flaw by installing a secondary solenoid valve and the new electronic logic is a normally closed fuel valve. A mature company could have survived a disaster, while a young company like Capstone probably could not have, so luck was with the company at that instance.

General Electric supposedly has control of the Honeywell microturbine design as part of the dissolution of the merger talks that occurred between those two companies in 2002. GE is purported to be looking at coming into the microturbine market either with a modified Honeywell design, a new design, or an acquisition of an existing microturbine company. Certainly Capstone, the only stand-alone microturbine company presently in the marketplace, with significant cash in the bank, an enviable history of installations and 4,000,000 combined hours of field operation would be a perfect target for a company such as GE to investigate.

The last chapter of microturbine evolution has yet to be written.

Appendices

APPENDIX A:
CASE HISTORIES

Included here are actual operating microturbines in a variety of applications which demonstrates the versatility and markets in which these machines can be used.

1. Plastics Manufacturing Plant
2. Metal Plating Facility
3. Campus Swimming Pool
4. Hotel
5. Supermarket
6. Waste Water Treatment Plant
7. Community Pool
8. Coal mine
9. Landfill Methane Recovery
10. Natural Gas Liquefaction Plant
11. Administration Building

APPENDIX B.
INSTALLATION STANDARDS FOR MICROTURBINES

The report on U.S. Installation, Operation and Performance Standards for Microturbines was prepared under the auspices of the DOE for a comprehensive review of codes and standards applicable to the safe installation and operation of microturbines in commercial and residential buildings.

APPENDIX C.
MICROTURBINE POWER CONVERSION TECHNOLOGY REVIEW

In this study, the Oak Ridge National Laboratory (ORNL) is performing a technology review to assess the market for commercially available power electronic converters that can be used to connect microturbines to either the electric grid or local loads.

Appendix A

Case Histories

CASE STUDY #1.
HARBEC PLASTICS, ONTARIO, NY

Quick Facts
Number of Units:
  25 Model C-30, 30 kWh microturbines
  Electrical Output - 750 kW
  Heat Output -1.5 MW

Additional Equipment:
  Power Server Device to Monitor Electrical Load
  Air to Water Heat Exchangers
  Absorber Chiller

Installation Purpose:
  To provide reliable electric power and eliminate outage costs
  To provide thermal output to drive heating and cooling systems
  To deliver better total fuel efficiency
  To reduce environmental impact

Results:
  * Yields 36% net energy cost reduction
  * Produces a simple payback of 2.5 years.

YEARS OF INCREASINGLY FREQUENT power problems culminated in a
string of grid outages and other power events that hit this manufacturer
hard in 1999. Something had to be done. Since even a momentary outage
was as damaging as a lengthy one, standby generation alone was not an
adequate solution. A battery UPS my have improved power reliability
but would have increased the already high electric costs. The solution

also needed to instantaneously follow the load to meet the facility's varying needs. Often, 30% swings in power needs would occur. The onsite system must be a continuous source of quality power to the facility with some level of redundancy.

Since Harbec is an ISO 14001-certified facility, Harbec's president was committed to designing an environmentally benign onsite system. This commitment led away from reciprocating engine technologies, even those fired by natural gas, as NO_x emissions from reciprocating engines are many times higher than the best-in-class generating technologies. Thus Harbec investigated various technologies including wind power, fuel cells and microturbines.

Harbec wanted a solution that recovered as much of the input energy as possible. They envisioned a system that would capture exhaust heat and put it to valuable use for space heating, process drying and power free air conditioning via absorber chilling. While attracted to the renewable energy supply represented by wind and the potential high efficiency and low emissions of fuel cell prototypes, economic analysis clearly revealed that microturbines, offered the best up-front and ongoing value for clean, continuous onsite power. The result was an array of 30 kWh microturbines with air to water heat exchangers and hot water fired absorber chiller. The array of microturbines and heat exchangers is shown in Figure A-1. Figure A-2 shows the absorber chiller.

To Generate Chilled Water for Air Conditioning
In this array, 25 microtubines were installed, 20 of which were

Figure A-1. Microturbines and Heat Exchangers at Harbec

Hot exhaust from the Capstone MicroTurbines fires this Carrier absorption chiller to air condition the Harbec facility using near-zero electrical load.

Figure A-2. Absorber Chiller Taking Hot Water from Microturbines

mated to five heat recovery exchangers. Water is heated to 210°F. Five of the microturbines provide redundancy for electrical production. In the winter, the hot water is routed to the ventilations system as well as radiant floor heating built into the concrete slab of the facility's warehouse area. In warm and humid weather, the heated water fires a 200 ton absorption chiller to air condition and dehumidify Harbec's warehouse and production areas without adding any appreciable electrical load. Humidity control also eliminates the cost of drying raw materials in humid weather.

This system went into full operation during the summer of 2001. The microturbine array provided full-time operating power to all facility loads, including the starting and stopping of several motors as large as 60 horsepower, while the exhaust heat fired the Carrier absorber chiller delivering a constant supply of dry chilled air to the facility.

These microturbines provide uninterrupted power to Harbec's fa-

cility and follow the load needs for many processes and machines within
the facility. Power outage costs and downtime are eliminated with these
Dual Mode units. Exhaust is clean with less than 9 ppm NO_x being
emitted, which is less than half that per kWh emitted by generators of
the local utility. Maintenance on the system is expected to be minimal.
After a full year of constant operation, the only scheduled maintenance
is an air filter change and a quick visual inspection of key system components.

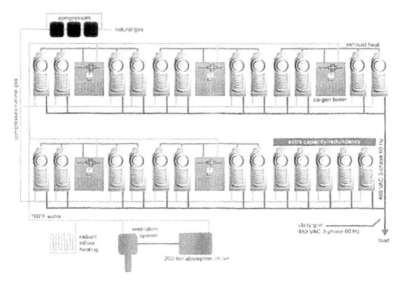

Anatomy of a Solution: A diagram of the system at Harbec Plastics.

Figure A-3.

CASE STUDY #2.
FAITH PLATING, HOLLYWOOD, CA

Quick Facts:
Number of Microturbines -   Four (4) 30 kWh, natural gas fired units
                               Electrical Output - 112 kWh @ ISO
                               Heat Recovery - 240 kWh or 845,000 Btu/hr

Additional Equipment -
     One (1) Air to Water Heat Exchanger

Installation Purpose -
　　To offset grid electricity usage and cost
　　Improve power reliability
　　To meet stringent emerging and proposed boiler NO_x standards
　　To derive value-added benefits of direct exhaust drying of sludge

Results -
　　Annual estimated savings of over $57,000
　　Projected 4-year simple payback
　　Reduction in waste weight and volume as well as savings in drying
costs

Faith Plating began operations in 1918 in Hollywood, CA. With movies, came cars and with the most popular models came chrome plated, ornately adorned bumpers and accessories. Faith Plating emerged as a leader in vehicle related plating.

Figure A-4. Chrome Plated Vehicle Accessories

Essential to Faith Plating's operations has been the need for a reliable supply of electricity and hot water. External factors threatened the business in 200 1:

*   Power shortages racked California, triggering blackouts and higher costs

*   The EPA issued new emission restrictions on hot water boilers

The Assistant General Manager of Faith wrestled with these issues and decided to take action by installing four (4) 30 kWh microturbines which would mitigate the high cost of electricity as well as providing power during grid outages. He had heard about microturbines while driving to work one morning from a radio broadcast. Similarly, the creation of hot water for the plating tanks would reduce the natural gas required for that process and qualify the facility for the attractive energy rebates the State of California was offering for cogeneration systems. The low emissions of the microturbines reduced Faith overall NO_x emissions to comply with the stiffer EPA requirements. The savings in emission controls that would have been required on Faith's boilers are not accounted for in the overall savings.

Figure A-5. Two of the four microturbines providing electricity and hot water to the facility. Note the ductwork providing clean, outside air to the inlet of the turbines.

The Dual Mode capability of the microturbines; allow switching between grid-parallel and stand-alone operation in the event of a grid outage. The microturbines have black start capability which enables them to start up even if they had been down when a grid failure occurred.

All of Faith's plating tanks were converted from conventional boiler operation to the new heating loop from the air to water heat exchanger installed as part of the cogeneration project. The loop runs at 170°F to 190°F. Each 14,000 gallon master tank holds plating solution, liquid metal in a sulfuric solution. Nickel, copper and chrome plating tanks must be maintained between 110°F and 190°F depending on the metal. Thermostatic controls on the tanks monitor proper temperature levels. Like other continuous process manufacturers, Faith Plating is reliant on dependable electrical power. Grid interruptions mean ruined work-in-progress along with costly downtime to clean and restart processes. The four microturbines provide 60% of Faith's needs during peak periods, sufficient to keep the facility's critical areas in operation during a grid failure and providing a slashing of utility bills of about $5,500 per month on electricity and a net average of $1,500 per month for the gas bill (after accounting for the natural gas burned to operate the turbines).

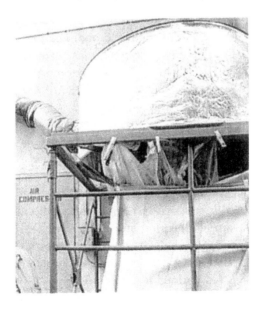

Figure A-6. Drying of Sludge Waste via Waste Heat

A further savings accrued by using the exhaust flow from the air to water heat exchanger to dry the waste sludge material collected from the plating tanks. By reducing the weight and volume of the sludge, the cost of removal was reduced both by eliminating the electric dying previously employed and the cost of sludge disposal.

So, what started as a radio blurb ended up in a major cost reduction program for Faith Plating by utilizing microturbines and cogeneration.

CASE STUDY #3.
CAMPUS SWIMMING POOL, SOUTHERN ALBERTA
INSTITUTE OF TECHNOLOGY

Quick Facts:
Number of Microturbines -
    Two (2) 30 kWh with integral heat recovery
    Electrical Output - 60 kWh @ ISO
    Heat Recovery - 120 kW or 422,280 Btu/hr.

Installation Purpose -
    To serve hot water needs of an Olympic sized swimming pool
    To generate onsite electrical power
    To reduce emissions from the traditional heating sources
    To provide SAIT students a hands-on learning opportunity
        in distributed generation and cogeneration

Results -
    Reduced energy costs
    An estimated 55% reduction in CO_2 emissions
    An estimated 97% reduction in NO_x emissions

For over 90 years the Southern Alberta Institute of Technology has built an international reputation in technical education from World War I era courses in steam engineering and motor mechanics to today's diverse curriculum encompassing engineering, information technology, health sciences, applied arts, business administration and apprenticeship trades. SAIT showcases various technologies giving students hands-on training with innovative systems.

Since SAIT was heating its Olympic sized pool with conventional hot water heating equipment and its electricity to run the pumps and other electrical needs was purchased from the local utility, a cogeneration system would seem to fill the economic as well as the intrinsic needs of the institution, i.e. teaching. The Institute already has a curriculum known as epicenter, a Center of Technology Development that trains the next generation of power engineers. Cogeneration would fit nicely into that curriculum.

Two (2) 30 kWh units, each with integral waste heat exchangers now supply the 60 nominal kilowatts of electrical energy while supplying over 4.22 therms/hour of heat energy for the pool.

The local microturbine distributor, Mariah Energy, developed a secure web-based dispatch, control and monitoring gateway via the microturbine's open communication protocols. Consequently, the "triple gate" supervisory system provides continuous data and enables Mariah to operate the systems in thermal load following, i.e. modulating power output to maintain constant temperature in the hot water loop. A secondary controller operates a mixing valve to regulate pool water temperature.

Saving money, saving the environment and helping educate the technology professionals of the future, this microturbine installation achieves many important goals simultaneously.

Overall thermal efficiency is over 80%. The recovered waste *heat is* used to heat the million liter pool as well as serve the domestic hot water

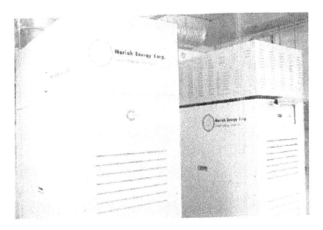

Figure A-7. Two (2) 30 kW Microturbines with Integral Air to Water Heat Exchangers

Figure A-8. SAIT's Million Liter Swimming Pool Heated by Cogeneration

needs of food services, showers and a Zamboni room The electricity goes to the Campus Centre, which houses a hockey rink, sport courts and a weight room offsetting energy SAIT would otherwise have to buy from the grid.

Although the purpose of the project was to reduce energy costs, reduction of cost volatility was a key component of the decision according to officials at SAIT. Because approximately 65% of the fuel used in the system would have been burned in SAIT's boilers, the value of the power is effectively 65% hedged against future fuel price fluctuations.

CASE STUDY #4.
HOLIDAY INN HOTEL, FARGO, ND

Quick Facts:
Number of Microturbines -
    One (1) 30 kilowatt unit
    Electrical Output - 30 kWh @ ISO
    Heat Recovery - 42 kW or 142,000 Btu/hr.

Installation Purpose -

To augment existing boilers with heat output to accommodate peak hot water demand

To generate electricity for onsite use

To provide data for EPRI/NRECA/DOE Distributed Resource Group

Results -

High reliability and system compatibility proven for hotel operations

Hot water availability boosted by 20%

More than 13,000 hours of near-continuous operation

Ability to operate critical electrical needs in the even of a grid outage

Hot water needs drive the majority of cogeneration projects and what better place to implement that technology in facilities that use hot water for guest rooms, laundries, kitchens, pools and spas. In other

The Holiday Inn in Fargo, NC, where Capstone solved a hot water shortfall while generating an electricity dividend.

Figure A-9. Hotel in Fargo, ND Using Cogeneration

words, a hotel. Most hotels already use natural gas for the heating process of water, but there is a way to improve fuel efficiency and the bottom line by turning that natural gas into more than just hot water—cogenerating electricity and hot water.

Gas fired boilers supplied hot water to 200 rooms in this hotel's two story wing. The system was overtaxed during peak accommodation periods and during these occupancy spikes the hotel had to also address the hot water needs of the laundry and kitchen. It appeared that rather than just install another boiler, to investigate cogeneration as a way to mitigate electrical costs while supplementing hot water needs. Since hot water was the priority, it was determined that the installation of a 22 kW AO Smith electric water heater would be the optimum solution. The electricity generated by the microturbine would run the electric water heater when it was needed and any excess electricity would be distributed throughout the hotel via the main electric panel.

The clean exhaust heat generated by the microturbine was looped into the hotel's commercial grad boiler via the air to water waste heat

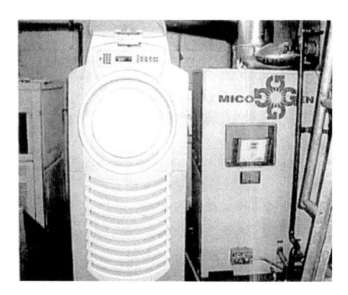

Clean exhaust heat generated by the Capstone MicroTurbine loops into the Holiday Inn's commercial grade boiler via a heat exchanger.

Figure A-10. Kilowatt Microturbine with Air to Water Heat Exchanger

exchanger and applied to the 1000 gallon hot water storage tank serving the laundry and kitchen. With simple site requirements, low noise output, small footprint and zero vibration, the microturbine system was sited near the laundry/kitchen storage unit.

Other than maintenance on the system's internal gas compressor, the system has run smoothly for more than 13,000 hours as of mid 2003. The dual mode controller allows the microtubine to operate in the event of a power outage by the grid supplying electricity to critical hotel areas such as front desk, lighting, reservation system and guest hallways. The internal gas booster compressor maintenance can be eliminated by implementing the manufacturer's foil gas bearing unit as a retrofit.

CASE STUDY #5.
SUPERMARKET - TOKYO, JAPAN

Quick Facts:

Number of Microturbines -
    One (1) 30 kilowatt unit
    One (1) liquid based desiccant dehumidification system
    One (1) 480v/220v transformer
    Electrical output - 30 kW
    Heat Recovery - 55 kW or 188,000 Btu/hr

Installation Purpose -
    To lower interior humidity and increase perishables' shelf life
    To generate electricity for peak shaving for power cost savings
    To drive desiccant system with waste heat from the microturbine
    To reduce greenhouse gas and pollutant emissions

Results -
    Successfully lowered store humidity to extend shelf life of perishables which constitute 80% of the market's revenue
    Regenerates desiccant system with exhaust heat of microturbine engine
    Lowers fees paid to electric utility
    Overall energy efficiency of 70% is attained

Two factors loom large at every supermarket: a large inventory of perishable goods and a large need for electricity to stock and sell those goods. Food quality is critical to a grocer's carefully balanced profit margins. The very air inside a market plays a key role. Maintaining the right humidity level extends shelf life, satisfying customers and the bottom line.

Similarly, because refrigerators and freezers are fundamental to food quality, a supermarket cannot reduce refrigeration needs to cope with higher peak power costs. However, it can choose a more affordable, supplemental electricity source when grid power is at a premium.

The Seiyu Supermarket enhanced their product shelf life while reducing their power costs by employing two distinct technologies: 1) the microturbine and 2) a Meidensha MioSpectruni desiccant de-humidification system. The microturbine supplies electric power to the store while its exhaust heat produces hot water for the desiccant dehumidifier. An internal air to water heat exchanger drives a hot water based desiccant dehumidification system manufactured by Tohoku Earth Clean. This system is integrated with the supermarket's air conditioning unit and lowers the in-store humidity to maintain maximum freshness of Seiyu's perishables.

The hot water exiting the desiccant system enters the Meidensha unit at 176°F. The water is sent back to the desiccant system at 194°F. In a representative cycle ambient relative humidity was 58%, with ambient temperature of 82°F. Air exiting the desiccant dehumidifier for store circulation measured 20% relative humidity and 85°F at a volume of 68,000 cubic feet per hour.

Since the microturbine's output power is 480 volts, a step down transformer was installed to meet the store's 220 volt electric panel through which the generated electricity was distributed. Since Japan's peak power rates escalate to 400 to 500% of normal rates, the microturbine system displaces a portion of that power.

Meidensha developed the MioSpectrum system, an integrated combined heat and power generator based on the 30 kilowatt microturbine. An OEM relationship exists between the microturbine manufacturer and Meidensha and is partly responsible for the over 300 microturbines currently operating in Japan.

These foods are preserved and presented thanks to Capstone. Hot exhaust dehimidifies the air to improve shelf life. Onsite electricity powers compressors and motors for refrigerators and freezers.

Figure A-11. Combination Air Conditioning and Desiccant Dehumidification System

This air handling unit sends reduced-humidity air into the Seiyu store after it has passed through a desiccant dryer that is regenerated by the Capstone MicroTurbine exhaust heat.

Figure A-12. Combination Air Conditioning and Desiccant Dehumidification System

Figure A-13. The Microturbine is the Core of the Meidensha MioSpectrum System

CASE STUDY #6.
WASTE WATER TREATMENT PLANT - CARDIFF, CA

Quick Facts:
Number of Microturbines -
    Three (3) 30 kWh units
    One (1) Air to Water Heat Exchanger
    One (1) Fuel Gas Booster Compressor
    One (1) Set of Gas Pre-Treatment Equipment
    Electrical Output - 90 kWh @ ISO
    Heat Recovery - 205 kWh or 700,000 Btu/hr.

Installation Purpose -
    To reduce power costs
    To provide heat to maintain temperature needs of digester
    To utilize the waste digester gas as a "free" fuel
    To reduce emissions formed during biogas combustion

Results -
    By utilizing the "free" fuel produced as part of the digestion process over 15% of the facilities electricity usage is produced.

Emissions of NO_x and methane were reduced vs. standard flaring practice

Provided rapid Return on Investment

Qualified for State grant money by producing clean, onsite power via cogeneration

In December of 2000, the San Elijo Joint Power Authority Water Reclamation Facility saw energy costs triple in a single month. This led to the investigation of onsite power production by utilizing the waste methane gas that is produced as a byproduct of anaerobic digestion. While reciprocating engines were proven in this field, the Facility felt that the higher cost of maintenance for that technology indicated looking for a better way to proceed. Microturbines provided that better way. The lower emission levels of microturbines were a plus.

Digesters use heated tanks that use microorganisms to break down waste solids and create biogas. San Elijo had been flaring this biogas and since the primary component of this gas is methane it can be readily used as a fuel. Allowing any methane to escape into the atmosphere can be ruinous as methane is 20 times as destructive as carbon dioxide as a greenhouse gas.

Biogas is generally wet and potentially has contaminants that can pose a problem to fuel gas booster compressors as well as the generators themselves, be they reciprocating engines or microturbines. Hence, adequate clean up equipment is needed to remove moisture and contaminants such as siloxanes that are commonly found in digester gas. Siloxanes are compounds of silica that find their way into sewage via discarded cosmetic products and other commonly used household products. Those silica compounds, when exposed to the heat of combustion in an engine using this gas for fuel, can form glass like coatings on the internal components of the engine. Over time, those coatings will drastically impede the performance of the engine—reciprocating or microturbine. Other contaminants such as hydrogen sulfide are tolerated by the microturbine up to 7% by volume and pose no operational problems.

Three (3) 30 kWh microturbine units were chosen for the job as an optimum match for the biogas being generated, the power generated and for the heat needed in the digesters themselves.

Since the biogas is generated at low pressure, basically atmospheric, and the turbines need 70 to 90 psig combustion chamber pressure, fuel gas boosters are needed to compress the digester gas to

San Elijo's three Capstone C30 MicroTurbines with heat exchanger. Homes surround the facility, confirming its responsibility to the community and the environment.

Figure A-14. San Elijo Waste Water Treatment Plant Microturbines

The Capstone MicroTurbine system runs on biogas, created when San Elijo's digester tanks process waste solids.

Figure A-15. Fuel Gas Booster Compressors at San Elijo Waste Water Treatment Plant

operating pressure. Normally scroll or sliding vane type compressors are utilized for this service.

The facility is experiencing savings in electricity costs of $4,000 to $5,000 per month according to the Facility's plant manager with a payback of capital costs, including the clean up gas treatment equipment, of three to four years.

CASE STUDY #7.
EUROPE'S LONGEST RUNNING MICROTURBINE -
PUTTEN, NETHERLANDS

Quick Facts:
Number of Microturbines -

One (1) 30 kWh unit

One (1) Air to Water Heat Exchanger

Electrical Output - 30 kWh @ ISO

Heat Recovery - 74 kW or 253,000 Btu/hr.

Installation Purpose -

To reduce recreation center energy costs

To create a more flexible and efficient pool heating system

To lower on site emissions

Results -

A 96% energy efficiency was attained

30% energy cost savings

Over 13,000 hours of runtime including one full year of uninterrupted service

Actual NO_x emissions of less than 3ppm

It is a basic law of thermodynamics that fuel will not completely convert to energy. There will always be losses and inefficiencies. Inefficiency is also an important economic concept. Inefficient use of a core commodity hurts the bottom line. In power generation, there is an intersection of thermodynamics and economics: a system's energy efficiency rating versus return on investment and profitability. The ratios become more favorable as efficiency increases. Typically, conventional generator systems yield energy efficiency from 25 to 40%, meaning substantially less than half of the purchased fuel becomes usable energy. Improving that ratio requires something beyond a conventional approach. Utilizing Combine Heat & Power, i.e. cogeneration typically achieves 70 to 90% total system efficiency.

However, Bosbad, a 400,000-gallon community pool in Putten, Netherlands, did not stop there. An incredible 96% total system efficiency was achieved!

The Netherlands natural gas utility, Gasunie, identified the Bosbad facility as an ideal fit for a microturbine system. The large, indoor/outdoor pool previously employed one central boiler. Gasunie proposed a cascade system: a 30 kWh microturbine functioning as a baseload unit—"boiler number one"—with eight (8), small, high efficiency boilers to supplement higher winter heating needs. With heat as the primary focus, the microturbine's electricity production would be used for loads such as pumps and lighting, the divident that would reduce the facility's total energy expenditures.

Figure A-. The Bosbad Community Pool in Putten, Netherlands

In winter months when the facility's indoor pool is in use, the "winter loop" which includes the secondary boilers and a beat exchanger, is periodically activated. In heating the indoor pool and the building housing it, the microturbine's energy efficiency is 87% as measured by Gasunie. During the summer when the outdoor pool is in operation, the microturbine is the sole boiler with its thermal output directly heating the water. During this season, energy efficiency is a record-setting 96% per Gasunie verification. Overall, the Bosbad facility has achieved a 30% reduction in energy costs.

Emission levels were measured on site for NO_x production and it was found that while the manufacturer guaranteed 9ppm of NO_x, actual emissions measure less than 3 ppm.

A further milestone was reached when the Bosbad microturbine became Europe's longest running microturbine with more than 13,000 hours of operation punctuated by a one-year block of uninterrupted runtime. This is typical of microturbine required maintenance, i.e. filter changes after—9,000 hours of operation.

The Capstone C30 MicroTurbine works both as a boiler and onsite electricity generator, converting 96% of incoming natural gas fuel into usable energy.

Figure A-17. The Microturbine and Heat Exchanger Setup at Bosbad

CASE STUDY #8.
A UNIQUE GREENHOUSE GAS SOLUTION -
SUMITOMO COAL MINING

Quick Facts:
Number of Microturbines -
    Five (5) 30 kWh Units
    Methane Collection System
    Ductwork and Blowers
    Electricity Output - 150 kW @ ISO
    Exhaust Recovery - 3.4 lb/sec mass flow

Installation Purpose -
    To cut greenhouse gas emission via methane combustion and CO_2 sequestration
    To demonstrate coal mine methane recovery technology
    To test CO_2 sequestration via exhaust injection

Results -
    Highly effective elimination of coal mine methane
    Hot exhaust gases are driven back into mine liberating more methane for power generation
    Onsite electrical generation powers facility loads and generates revenue
    Low maintenance and low emissions

For half a century, the Sumitomo mine in Akabira, Japan, supplied coal to a burgeoning Japanese economy, reaching an ultimate depth of 3,900 feet to produce nearly 80 million tons of coal. The mine was closed in 1994 as more natural gas was used to fuel Japan's industry.

However, the abandoned mine continues to discharge coal mine methane (CMM), a highly potent greenhouse gas 20 times more destructive than CO_2. CMM also presents explosion hazards, demanding constant management and dispersal. According to Sumitomo officials, Akabira released nearly 3.5 billion cubic meters of methane prior to its closure. An estimated 900 million remain. Even though production has ceased, Sumitomo knew their responsibility did not.

Flaring is a common practice to burn off methane emissions from mines, oil wells, landfills and wastewater treatment plants. Around the

Sumitomo closed the Akabira mine in 1994, but potent greenhouse gases remained.

Figure A-18. The Akibira Coal Mine

Exhaust from the five C30s is sequestered into coal seams to liberate more methane for power production.

Figure A-19.

globe scores of such sites have switched from flaring to combusting their waste methane with microturbines, which eliminates the gas with far fewer emissions than flaring. The waste gas becomes fuel for the microturbine system, eliminating the methane far more effectively and creating two other valuable dividends—creation of electricity and heat without using any commercial fuel. In other words—cogeneration.

Sumitomo created a "closed loop" system that eliminates the maximum amount of coal mine gas with the lowest possible emissions of methane, carbon dioxide and nitrogen oxides.

The five (5) microturbine array takes in 30% of the mine's total methane discharge.

Electricity from the microturbines powers a series of pumps and blowers that extracts methane gas from the mine. Surplus power is exported to a nearby factory. The methane travels to a pumping and compression plant before going to the microturbines. Exhaust from the five microturbines is collected in a common manifold and driven down the shaft via a dedicated system of fans and ducts. Carbon emissions from the exhaust are permanently sequestered in remaining coal seams, releasing more methane as the mineral pores preferentially take up the CO_2. This is the final step in the "closed loop" process, liberating more methane to maintain the continuum.

Sumitomo Coal Mining is considering this "closed loop" approach for other locations, multiplying the possible revenue and tangible environmental benefits on a wider scale.

CASE STUDY #9.
BIOGAS FROM A PUBLIC LANDFILL -
LOPEZ CANYON, LOS ANGELES, CA

Quick Facts:
Number of Microturbines -
Fifty (50) 30 kW units
Gas Compression, Drying and siloxane removal system
Electricity Output - 1.3 MW net

Installation Purpose -
To combust landfill gas more thoroughly with lower NO_x emissions than flaring would produce

To generate electricity from a renewable fuel source

Results -

Generates up to 1.3 MW of electricity from renewable "waste fuel"

Reduces annual NO_x emissions by more than 10,000 pounds

Destroys methane, a powerful greenhouse gas that is 20 times more damaging than CO_2

Lower maintenance needs compared to conventional biogas fueled engines

Lopez Canyon landfill was closed in 1996 but was still discharging methane from the decades of layered trash buried and decomposing there. A collection system was capturing this methane and other toxic gases and directing the gas to a flare system that was resulting in many tons of NO_x and other emissions being released to the atmosphere.

In a unique cooperative partnership between the Los Angeles Department of Water and Power (LADWP) and the South Coast Air Quality Management District (SCAQMD), fifty (50) microturbines were installed at the landfill to use the generated methane as a fuel to produce 1.3 MW of electricity which LADWP distributes throughout its marketing area.

Landfill gas is low in methane content—about 300-400 Btu/cubic foot, compared to 1,000 Btu/cubic foot for natural gas. Further, it has moisture and contaminants in the gas that require significant gas pre-treatment before going to the microturbines for combustion. Moisture removal is accomplished via refrigerated air dryers chilling the gas to 40°F and condensate removal systems, while the more severe contaminants such as siloxanes are removed via carbon granule filtration. After moisture removal the gas is heated to 18°F above its dew point to prevent any further moisture condensation. All other gaseous components are destroyed in the microturbine's combustion chamber.

Microturbines are proving to be a practical and cost-effective technology for biogas applications from landfills to anaerobic digesters at waste water treatment plants and agricultural/livestock facilities. Microturbines offer vastly reduced scheduled maintenance compared to conventional reciprocating engine driven generators, they are designed to operate 24 hours per day, 7 days per week at full load and produce 95% less contaminants to the atmosphere. This latter statistic is what won particular favor for the SCAQMD, the world's most stringent air quality management organization.

The LADWP's Green LA Program won the "Renewable Company of the Year" Global Energy Award from *Financial Times*, an award attributed in part to the use of Capstone MicroTurbines at Lopez Canyon.

Figure A-20. Installation of Fifty (50) 30 kWh Microturbines at Lopez Canyon Landfill

CASE STUDY #10.
NATURAL GAS LIQUEFACTION PLANT, MINNEAPOLIS, MN

Quick Facts:
Number of Microturbines -
  Two (2) Units, one 60 kW unit and one 30 kW unit
  One (1) Air to Water Heat Exchanger
  One (1) Desiccant Wheel Dehumidifier
  Electrical Output - 90 kW @ ISO
  Heat Recovery - 58 kW or 200,000 Btu/hr from 30 kW unit.

Installation Purpose -
  To reduce electric grid demand through peak shaving
  To create non-electric air conditioning through CHP

Results -
 Produced more than 19,000 hours of highly reliable service
 Reduced energy cost by offsetting grid purchased power
 Reduced maintenance cost versus other generator technologies
 Reduced emissions versus other generator technologies

Located 10 miles south of Minneapolis, Minnegasco's Dakota Station liquefies and stores natural gas during spring and summer off-peak months when commodity gas prices are lower. The station buys electricity under an energy and demand rate schedule. The 30 kW unit, with heat recovery and desiccant dehumidification, produces space beating in the winter and reduces cooling needs in the summer via the desiccant wheel by removing moisture from the air. The 30 kWh unit was installed in load following, grid connect mode and produces virtually all the electrical needs for the facility during three quarters of the year.

The 60 kWh unit was brought on line to provide standby power for the station's propane refrigeration system, keeping 6,000,000 gallons of propane at -50°F. This unit has logged more than 8,000 hours of runtime.

Dedicated to providing economical, dependable energy for customers, Dakota Station uses Capstone MicroTurbines for the site's own economical, dependable energy solutions.

Figure A-21. The Minnegasco's Gas Liquefaction Plant in Minnesota

The 60 kW unit employs the "Time of Use" feature which programs the microturbine to run during the day when electric rates are higher and to shut down at night when the rates are lower.

Management at the plant comment that this system is an amazing combination of great design, good economics, high reliability and environmental sensibility. It certainly assists in the bottom line of controlling our costs of operation. The public expects Minnegasco to stand for energy efficiency and we put our technology where our mouth is.

CASE STUDY #11.
UNINTERRUPTED TELECOMMUNICATIONS, HOUSTON, TX

Number of Microturbines -
    One (1) 30 kW Unit
    Electrical Output - 30 kWh @ ISO

Installation Purpose -
    To protect against grid outages

Results -
    Ensures non-stop data network/telecom operation via 24/7 onsite power
    Small footprint and sound signature
    Minimal scheduled maintenance

Mandating an alternate source of power support for its data/ telecom system, Boeing Co. of Houston, Texas required 20 kilowatts of power for six hours duration. UPS batteries and genset engines were the conventional choice, but not the best fit with the company's requirements. Emissions and vibration, as well as noise, would be difficult to mitigate.

Opting for a different power solution Boeing chose to install a 30 kW microturbine in a pre-existing concrete enclosure near the building's loading dock. The microturbine showed its versatility in being able to be installed in a close quartered situation.

The data/telecom system utilizes dual power inputs: one from the microturbine and one from the grid. In the event of a grid power outage, the microturbine ensures uninterrupted service. If the microturbine goes

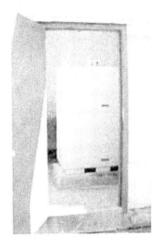

Doorway: The Capstone C30 MicroTurbines was placed in an existing structure at the Boeing site, making deployment simple.

Figure A-22. Microturbine Installed in Small Room

off-line, the utility grid is the backup source. Boeing required no special emissions or hazardous materials permitting thanks to the microturbine's ultra low emissions and air cooled/air bearing design. The small footprint and lack of vibration allowed the installation in the small room.

Appendix B

Installation Standards for Microturbines

U.S. INSTALLATION, OPERATION, AND
PERFORMANCE STANDARDS FOR
MICROTURBINE GENERATOR SETS

A.-M. Borbely-Bartis

J.G. DeSteese

S. Somasundaram

August 2000
Prepared for the U.S. Department of Energy
under Contract DE-AC06-76RL01830
Pacific Northwest National Laboratory
Richland, Washington 99352

EXECUTIVE SUMMARY

As the nature of energy and electricity demand changes in the post-internet economy, demand for a variety of distributed energy resources (DER) grows—natural-gas fired microturbines are just one of a

suite of new power generation products entering (or soon to be) the
U.S. marketplace. Developing an environmentally benign, efficient and
cost-effective technology is paramount for all DER developers. But
without the manufacturing, performance, installation and operation
codes & standards necessary to support that technology, deployment—
and ultimately financial success—may be delayed for years.

Microturbines are typically single shaft machines (one company
is developing a twin-shaft) with the compressor and turbine mounted
on the same shaft as the electrical generator. It therefore consists of
only one rotating part, eliminating the gearbox and associated moving
parts. This direct-drive, high-speed design has enabled manufacturers
to create a more reliable compact power plant than traditional engine
generator sets.

Codes and standards that regulate the safe performance of microturbine generators in or near buildings outside traditional utility jurisdiction will fall into one of three realms, and have been separated into
chapters within this report.

Manufacturing Standards involve the materials, design and construction of a single unit, commonly referred to in the C&S industry as
"the box." Microturbines do not currently have a manufacturing or
performance standard by which each unit can be evaluated.
Underwriter's Laboratories has listed one company's product under
UL 2200, "Stationary Engine Generator Assemblies." The standard is
under review for modifications to include microturbines by reference.

Installation and Operation Standards address electrical- and fire-
safety, and life-safety requirements such as emergency backup systems
for life-support operations or mission-critical systems. UL 2200 is most
relevant here, as is EGSA 101, NFPA 37, 101 or 110, IEEE 519 (harmonic distortion) or ASME B133.8 (noise emissions).

Interactions Between the MT Unit and Other Building Systems
include fire protection, egress, ventilation, electrical shock protection,
and fuel supply. The building code official will want to see sufficient
space around, access to, and protection of the MT unit to ensure public
safety under all conditions. Historically, wide-scale power generation
technologies have been owned and operated by regulated utilities and
have not been subject to local code requirements. However, in 2000,

the International Building Code declared that utility-operated facilities not owned by—and dedicated to—the utility will fall under local code jurisdiction. Currently, no technology-specific code structure exists for the installation and base load operation of microturbines by private nonutility parties in commercial or residential buildings.

In such cases when a technology is not specifically referenced in the applicable code documents, code officials may disallow the installation and operation of that unit. Building officials may require specific tests to demonstrate "alternative compliance"; performing the necessary site-specific studies—possibly requiring modifications to the installation—can increase the installed cost of a unit beyond any reasonable expectation for return on investment. Further, with 44,000 state and local code jurisdictions in the United States, the absence of any reference in the national code bodies may kill altogether the early prospects for the mass deployment of a new energy technology such as microturbines.

This report reviews the codes and standards applicable to the safe installation and operation of microturbines in commercial and residential buildings. It provides an overview of potential regulatory roadblocks, as well as recommendations for further action that may be undertaken by the U.S. microturbine community.

This report does not attempt to provide a technical review of any manufacturer's product, nor does it involve any technology-specific analysis (i.e., comparing the output current grounding design of one manufacturer, against UL 2200 requirements for Stationary Engine Generator Assemblies).

This study did not find that additional standards are required for building code officials to accept microturbine installations when external to a commercial or industrial facility. However, a general lack of education among code officials on the differences between DER installations and traditional backup generators, is in fact hampering the deployment of microturbines. A general education program for building code officials is in order. The report closes with the recommendation to pursue a performance standard specifically for microturbines, and for those applications most relevant to the emerging DG market—combined heat & power, premium power quality.

CONTENTS

1.0   INTRODUCTION

Codes and standards that regulate the safe performance of micro-turbine generators in or near buildings outside traditional utility jurisdiction will fall into one of three realms:

- manufacturing specifications and design requirements of the unit or system (i.e., "the box")

- installation and operation

- interactions between the microturbine unit and other building systems, structures, or life-safety issues.

Historically, wide-scale power-generation technologies have been owned and operated by regulated utilities and have not been subject to local code requirements. However, in 2000, the International Building Code declared that utility-operated facilities not owned by—and dedicated to—the utility will fall under local code jurisdiction. Currently, no technology-specific code structure exists for the installation and baseload operation of microturbines by private nonutility parties in commercial or residential buildings.

In such cases when a technology is not specifically referenced in the applicable code documents, code officials may disallow the installation and operation of that unit. Building officials may require specific tests to demonstrate "alternative compliance"; performing the necessary site-specific studies–possibly requiring modifications to the installation—can increase the installed cost of a unit beyond any reasonable expectation for return-on-investment. Further, with 44,000 state and local code jurisdictions in the United States, the absence of any reference in the national code bodies may kill altogether the early prospects for the deployment of a new technology.

1.1 Microturbine - Definitions

Microturbines are typically single-shaft machines (one company is developing a twin-shaft) with the compressor and turbine mounted on the same shaft as the electrical generator. It therefore consists of only one rotating part, eliminating the gearbox and associated moving parts. This direct drive high-speed design has enabled manufacturers to create a more reliable compact power plant than traditional engine generator sets.

The fuel of choice is natural gas delivered at pressures exceeding 55 psi (this requirement may go as high as 90 psi), although low-pressure gas can be boosted with centrifugal or scroll-type compressors. Systems can operate on propane as well, which requires on-site fuel storage.

Virtually all MTs are installed with recuperators to achieve 28-30% electrical efficiency. Unrecuperated MTs generally run at 14-17% efficiency (LHV). The recuperator is a heat exchanger, transferring heat from the exhaust gas to the discharge air. Before it enters the combustor, the exhaust gas is reduced to near compressor discharge temperature and the compressor discharge air is heated to near turbine exhaust gas temperature. The heat added to the air reduces the amount of fuel required to raise the temperature to that required by the turbine.

The electrical output is a high-frequency AC (1500-4000 Hz, 3-phase). The voltage is rectified and inverted to a normal 3-phase 50 or 60 Hz. In most systems, the power inverter and the alternator are used as the turbine starting system. In some cases two 12-volt batteries provide system power; those without "black start" capability (e.g., capable of starting independently) require AC power from the grid for their initial starting operation.

1.2 Methodology

This report was prepared by analyzing manufacturer's specifications against the most common industry sources for design, manufacturing, performance, installation and operation standards for electrical equipment installed in the United States. The following agencies provided primary or secondary information in this report:

American National Standards Institute (ANSI)
American Society of Heating, Refrigerating and Air-Conditioning Engineers, Inc. (ASHRAE)
American Society of Mechanical Engineers (ASME)
American Society for Testing and Materials (ASTM)
Electrical Generating Systems Association (EGSA)
Institute of Electrical and Electronics Engineers, Inc. (IEEE)
National Electrical Manufacturers' Association (NEMA)
National Evaluation Service (NES)
National Fire Protection Association (NFPA)
Underwriters Laboratories (UL)

2.0 CURRENT CODES AND STANDARDS

Building codes and manufacturing and installation standards were not developed with the intent of regulating the generation and distribution of electricity (considered to be a function of regulated utilities). Nor were the codes developed to regulate the generation and possible distribution of electricity by others. Insurance companies, lenders, building owners, tenants, occupants, and the general public expect buildings to be safe places. The code official is charged with protecting health and life-safety in buildings. He or she also is charged with enforcing codes as they currently exist. Because existing codes and standards were not written with specific requirements for generation and distribution of electricity by microturbine products, trying to apply them to technologies they were not intended to address places the enforcing agency (most likely a municipality) at risk for liability in a life- or property-threatening event.

Where there is no clear direction from codes and standards, there is potential for differing interpretations. These differing interpretations lead to differing installation requirements that can significantly impact both costs and length of time between purchase and operation. In evaluating a proposed installation design, particularly when submitted under the code provision "alternative methods and materials," a code official may require that the design undergo extensive engineering analysis and specialized testing before granting approval.

For the various codes and standards to formally mention DG products, the lag-time is approximately three years. Most national model codes operate under a three-year revised publication cycle. To avoid future code-related conflicts, changes proposed for a specific code must be fully integrated and coordinated with other codes. Where possible, code change proposals should address the full range of DG products. Uniform requirements for grid interconnection for all DG products would be preferable to requirements that vary by product. DG manufacturers, utilities, and others should work together on codes and standards issues to develop code change proposals for uniform requirements that will be adopted by the national model code agencies. The alternative could be a maze of requirements that serves as another barrier for design professionals, contractors, utilities, and code officials. Collaboration and cooperation can produce uniform code requirements that expedite the installation of DG products.

This chapter provides an overview of the U.S. standards and codes most likely to be relevant to microturbine technologies. Voluntary consensus standards are summarized in Section 2.1; model codes are discussed in Section 2.2.

2.1 Voluntary Consensus Standards

Voluntary consensus standards are documents that can be used or referenced to ensure uniformity in the testing, rating, evaluation, or design of products, materials, and other items. The term consensus is used to describe general agreement but not necessarily unanimity on a particular issue. Standards generally are referenced or transcribed within model codes. Standards also usually are focused on a specific aspect of the subject of interest.

2.1.1 Types of Standards

Product testing standards or test methods provide for the uniform testing and evaluation of a product or material. One example is a test to determine the combustion characteristics of various building materials (rate of flame spread, smoke emission profile, etc.). Although other documents establish the maximum acceptable values for these effects, a uniform test is necessary to determine these numbers irrespective of material, so different materials/products can be compared fairly.

Rating standards address the performance of a specific product or material. An example of a rating standard is one for air-source heat pumps developed by the Air Conditioning and Refrigeration Institute (ARI). Such a standard allows all such equipment to be tested and rated so the output at various input and ambient conditions can be measured and results equitably compared from manufacturer to manufacturer. Other examples in this category include tensile strength and/or flexibility of a stated material (concrete).

Minimum acceptable design or construction standards typically do not involve testing or rating a specific product, but rather the attributes required in a specific building design. For this reason the systems interacting within–and reacting to—that design may prove to be difficult to quantify. The American Society of Heating, Refrigerating and Air-Conditioning Engineers, Inc.(ASHRAE) Standard 90.1 consumed a decade of analysis, debate and open meetings before establishing criteria to represent an energy-efficient commercial building.

Voluntary consensus or industry standards are neither legislation nor regulations. However, these standards might be adopted into law by federal, state, or local agencies. Until adopted, they are as the name implies—voluntary. As new technologies, products, or processes are developed, new standards typically rise to cover testing, rating, and design.

Circular A119 of the U.S. Office of Management and Budget requires federal government agencies to use voluntary standards for regulatory and procurement purposes, when appropriate.

2.1.2 Standards Developers

The following paragraphs provide brief descriptions of organizations most involved in the development of standards related to the power generation industry. These organizations do not write the standards but provide the protocols and process support for standards development. While staff members provide the necessary support functions, the actual development and maintenance of a given standard and technical interaction with those who comment on that standard are performed by volunteers who represent a broad range of interests.

Although the American National Standards Institute (ANSI) coordinates standards initiatives within the United States, there are those who do not use ANSI processes. In addition, there are standards 2.3 initiatives in many other countries as well as at the international level. These may be known to ANSI, especially where the U.S. has some involvement in the international level activities. If information on standards development is desired, ANSI should be contacted first, followed by a contact to the likely standards generating organization, trade association representing the technology or issue in question, and manufacturers or others involved with the subject area. The National Institute of Standards and Technology (NIST) is another source of information on standards.

American National Standards Institute (ANSI)

ANSI is the dominant developer of consensus standards in the United States today, with over 13,000 consensus-based national standards in existence. ANSI is the sole U.S. representative of the two major non-treaty international standards organizations, the International Organization for Standardization (ISO), and, via the U.S. National Committee (USNC), the International Electrotechnical Commission (IEC). ANSI is also one of five permanent members to the governing ISO Council, and one of four permanent members of ISO's Technical Management Board. Through its ISO representation, ANSI has the authority to vest Technical Committees (TCs) in concert with international standards-making bodies.

U.S. standards are frequently presented for consideration (through ANSI) to the ISO or IEC where they are adopted in whole or in part as international standards.

American Society of Heating, Refrigerating and Air-Conditioning Engineers, Inc. (ASHRAE)

The American Society of Heating, Refrigerating and Air-Conditioning Engineers, Inc.(ASHRAE) is a 50,000-member international organization with chapters throughout the world. The society is organized for the sole purpose of advancing the arts and sciences of heating, ventilation, air conditioning and refrigeration for the public's benefit through research, standards writing, continuing education and publications.

Through its membership, ASHRAE writes standards that set uniform methods of testing and rating equipment and establish accepted practices for the heating, ventilation, air-conditioning, and refrigeration (HVAC&R) industry worldwide, such as the design of energy-efficient buildings. The Society's research program, currently more than 100 research projects worth nearly $8 million, investigates issues such as identifying environmentally benign refrigerants. ASHRAE organizes broad-based technical programs for presentation at its semiannual meetings and co-sponsors the International Air-Conditioning, Heating, Refrigerating Exposition, the largest HVAC&R trade show in North America.

ASHRAE writes standards and guidelines in its fields of expertise to guide industry in the delivery of goods and services to the public. ASHRAE standards and guidelines include uniform methods of testing for rating purposes, describe recommended practices in designing and installing equipment, and provide other information to guide the industry. ASHRAE has some 87 active standards and guideline project committees, addressing such broad areas as indoor air quality, thermal comfort, energy conservation in buildings, reducing refrigerant emissions, and the designation and safety classification of refrigerants.

Of most importance to the microturbine development community, ASHRAE Technical Committee 9.5 is currently studying the possibility of developing a standard for cogeneration systems. It is vital that the microturbine community be involved in this effort.

American Society of Mechanical Engineers (ASME)

The American Society of Mechanical Engineers (ASME International) has been involved in the development of codes, standards, and conformity assessment programs since 1884. The organization maintains and distributes 600 codes and standards used around the world for the design, manufacturing, and installation of mechanical devices.

ASME B133 is the ASTM-referenced standard for gas turbine procurement; included within this family of standards are subcommittees on fuels, performance, controls and auxiliary equipment, maintenance and reliability, and sound emissions. These standards were developed in the 1980s and have not been updated to include microturbines.

American Society for Testing and Materials (ASTM)

The American Society for Testing and Materials (ASTM) is a not-for-profit organization that provides a forum for standards development for materials, products, systems, and services. From the work of 132 standards-writing committees, the ASTM publishes standard test methods, specifications, practices, guides, classifications, and terminology. The ASTM's standards development activities encompass metals, paints, plastics, textiles, petroleum, construction, energy, the environment, consumer products, medical services and devices, computerized systems, electronics, and many other areas. ASTM Headquarters has no technical research or testing facilities; such work is done voluntarily by 35,000 ASTM members throughout the world.

More than 10,000 ASTM standards are published each year in the 72 volumes of the Annual Book of ASTM Standards. These standards and related information are sold throughout the world.

Electrical Generating Systems Association (EGSA)

The Electrical Generating Systems Association (EGSA) is the association for on-site power generation. Its members include manufacturers, dealer/distributors, manufacturers' representatives, consulting engineers, and others interested in the on-site power-generating industry and the components of electrical power-generating systems. The association develops educational materials, conducts seminars, publishes the bi-monthly magazine "Powerline," sponsors technical meetings, and develops national standards involving the use of on-site power. In addition to these inter-industry standards, the EGSA works to develop performance standards for equipment and components specific to its industry.

Institute of Electrical and Electronics Engineers, Inc. (IEEE)

The Institute of Electrical and Electronics Engineers, Inc. (IEEE) is a not-for-profit association and has more than 330,000 individual members in 150 countries. Through its technical publishing, conferences and consensus-based standards activities, the IEEE produces 30% of the world's published literature in electrical engineering, computers, and control technology, holds annually more than 300 major conferences, and has more than 800 active standards with 700 under development.

National Electrical Manufacturers' Association (NEMA)

The National Electrical Manufacturer's Association develops standards for electrical component or original equipment manufacturers. The organization publishes 200+ standards for

- building equipment
- diagnostic imaging and therapy systems
- electronics
- industrial automation
- industrial equipment
- insulating materials
- lighting equipment
- power equipment
- wire and cable products

NEMA standards are commonly cited in DOD regulations, the National Electrical Code, UL standards, and DOE standards for electric motors. NEMA is also a founding member of CANENA (Council for the Harmonization of Electrotechnical Standards of North America), a trinational organization devoted to harmonizing NAFTA-member standards for electrical equipment.

National Fire Protection Association (NFPA)

The National Fire Protection Association (NFPA) is an international, nonprofit, membership organization founded in 1896 to "protect people, their property and the environment from destructive fire."

The NFPA publishes the National Electrical Code, the Life Safety Code, the Fire Prevention Code, the National Fuel Gas Code, and the National Fire Alarm Code. The organization operates in 100 countries with 65,000 volunteers and staff.

NFPA codes most commonly referenced by local code officials include

- NFPA 1, Fire Prevention Code
- NFPA 13, Standard for the Installation of Sprinkler Systems
- NFPA 54, National Fuel Gas Code
- NFPA 58, Standard for the Storage and Handling of Liquefied Petroleum Gases
- NFPA 70, National Electrical Code
- NFPA 72, National Fire Alarm Code
- NFPA 101, Life Safety Code

Underwriters Laboratories (UL)

Underwriters Laboratories Inc. (UL) is an independent, not-for-profit product safety testing and certification organization. Founded in 1894, UL has five testing laboratories in the United States and subsidiaries in Mexico, Denmark, England, Italy, India, Singapore, Taiwan, Hong Kong and Japan. The organization also has numerous international, affiliate, and representative offices, as well as field representatives located throughout the world. Most significant for the microturbine community, UL 2200 is the product standard by which microturbines are currently being rated.

UL-Canada Mark

Through the provisions of agreements between UL and Canadian certification organizations, UL clients can receive UL and Canadian certifications with one submittal. These agreements provide for the reciprocal acceptance of test results and cover a wide range of products.

UL has been granted Certification Organization (CO) and Testing Organization (TO) accreditations for Canada by the Standards Council of Canada (SCC). UL's CO accreditation encompasses all of its facilities that handle both certification and testing, and all of its product categories and programs. By virtue of this accreditation, UL can evaluate products intended for the Canadian marketplace to Canadian National Standards and Codes, and authorize clients to label those products with the UL Mark for Canada.

2.1.3 Consensus Standard Development Process

A new technology, government directive, or other phenomenon may create the need to test, rate, or otherwise evaluate something that has not been addressed previously. An example is the development of ASHRAE Standard 90 dealing with building energy efficiency in 1975. This was a direct response to the 'Energy Crisis' and the need expressed by some states for criteria on energy conservation that could be incorporated into their building codes.

Once the need is identified (and a draft standard developed by some interested entity as a "strawman"), the individual or organization with the greatest desire for standardization will seek out a standards developer and request the developer to initiate standards development. In the case of ASHRAE, the organization's Standards Committee makes the decision to establish new 'Standards Project Committees.' Within the NFPA, the Standards Council makes the decisions on new projects.

The standards developer, typically with help from the standards instigator, will draft a title, purpose, and scope (TP&S)for the standard. These establish the framework for the standards project. This document, along with an expression of the need for the standard, is then considered by the committee that has such approval authority (Standards Committee, Standards Council, and so on). These committees may also consider conflicts with other existing standards and the scope of the project as it relates to the goals and objectives of their organization.

2.2   Model Codes

Model codes are just as the name implies—models of a code that anyone (typically federal, state, or local agencies) can adopt to address design, construction, and operation of buildings and facilities. Without a model code, each adopting entity would have to develop its own unique criteria—a costly, time-consuming, and easily outdated process. Model codes are written in enforceable language that meets the needs of the enforcement community. Model codes are written in a prescriptive manner so that the requirements are easily understood. Model codes usually present detailed prescriptive criteria and then allow alternative materials, methods or equipment based on performance equivalency.

They contain minimum requirements that are needed to protect the health, life-safety, and welfare of the public. Issues such as carpet quality would not be addressed within a model code, although the fire characteristics of the carpet would be addressed.

Model codes are not legislation or regulations, nor are they standards as described in Section 2.1., but when formally adopted, they do become enforceable law. Model codes rely on standards developed by others and refer to them as needed. A model code provides the regulatory framework under which a standard can be referenced.

Model codes offer an equal opportunity to all to participate in the development and revision process. Anyone can submit a code change; publication of code changes is available to all who request copies and open hearings are held, in which the merits of each proposed change are debated. The code officials vote on what they consider acceptable changes, based on oral and written testimony, and changes are finalized every year. This contrasts with the standards development process, in which it is possible to intervene in the process of development and publication through numerous comment and appeal processes.

Within each of the model codes, numerous subject areas are covered. And revisions to one subject area may impact (or be affected by) other areas. For example, studies have shown that energy could be conserved in some climatic locations if cathedral ceilings did not have to be ventilated. The building code mandates such ventilation and overrides the energy code. The use of photovoltaic shingles may be impacted by provisions in the building code for roof coverings. The mechanical code contains provisions for the safe installation of ground source heat pumps, which could adversely affect the deployment of that equipment. The fuel gas code has, in some instances, precluded the installation of high-efficiency gas heating equipment because it did not meet the venting requirements contained in the code.

The national model building codes generally incorporate by reference other model codes and standards as establishing the requirements for specific equipment installations or systems such as electrical, fuel gas, mechanical plumbing, and fire protection. Provisions of the building codes apply to the construction, alteration, moving, demolition, repair, maintenance, and use of any building or structure. State and local amendments may expand the scope of the national models to include such specific issues as emissions, air quality, noise levels, and other siting concerns.

Model codes are important because they form the basis for the vast majority of federal, state, and local building construction regulations in the U.S. New York state and city, Wisconsin, Chicago, and

Dade and Broward Counties in Florida are a few government entities that continue to write their own codes. There are over 44,000 jurisdictions (county, city) in the U.S. that could adopt and enforce codes. Even at the state level (about 25 states have statewide codes) writing and maintaining a building code is a time-consuming process. The model codes provide efficiency and uniformity for these potential "customers."

A code mandate to install specific devices (i.e., smoke detectors) can have a profound impact on product sales for a particular industry. Conversely, excessive installation and test criteria within a code, for certain products or equipment, can have the opposite negative impact on an industry.

2.2.1 Types of Model Codes

Building construction regulations provide the minimum requirements that a building and its systems, materials, and equipment must meet, and may vary by county, state or federal jurisdiction. The charter of building codes is to "safeguard life or limb, health (occasionally property), and public welfare by regulating and controlling the design, construction, quality of materials, use and occupancy location and maintenance of all buildings and structures and certain equipment specifically regulated" [Uniform Building Code (ICBO), International Conference of Building Officials (1997)].

2.2.2 Model Building Code Developers

Historically, model building codes were developed by three regional organizations outlined below: ICBO, BOCA, and SBCCI. The three regional model codes often had overlapping and contradictory building requirements, increasing the difficulties developers faced in managing production and equipment acquisition costs. In the 1994 the three model code groups agreed to work cooperatively within a single code organization to develop national building guidelines (see 'ICC' below).

As previously noted, the NFPA also develops documents that employ "code language" and as such could be classified as "model codes." Because these documents are developed by a standards organization, they are not thought of as model codes even though they are formatted, adopted, and used in much the same way as model codes. NFPA 70 and 54 (National Electric Code and National Fuel Gas code) are two examples.

International Building Code 1997 -
International Code Council (ICC)

In 1994 the model code groups formed the ICC and set out to develop one set of model codes. This family of international codes (I-codes) includes but is not limited to Energy Conservation, Fuel Gas, Mechanical, One- and Two-Family Dwelling, Plumbing, Building, Fire, and Residential. Currently the ICC is on schedule to publish in final form all ICC codes by 2000. However, adoption of the ICC codes by jurisdictions will be a slow process, because many jurisdictions must adhere to specific schedules, or must continue to use existing codes until there is justification for change. The regional model code groups will continue to operate. While cooperating within the ICC, they will continue to 'compete' in serving the building and code communities as they have in the past (through education, plan review, and other member services).

Uniform Building Code 1997 -
International Conference of Building Officials (ICBO)

The Uniform Building Code has been the dominant choice for adoption and enforcement by mid-western and western states including Indiana, Iowa, Minnesota, North Dakota, South Dakota, Montana, Wyoming, Colorado, New Mexico, Arizona, Utah, Idaho, Washington, Oregon, California, and Alaska. Provisions of the code shall apply to the construction, alteration, moving, demolition, repair, maintenance, and use of any building or structure. This building code references other codes for specific systems such as electrical, HVAC, plumbing, and fire protection. It does not specifically reference microturbines, nor place special restrictions on structures that might house or support microturbines.

National Building Code 1996 -
Building Officials and Code Administrators International (BOCA)

With the exception of New York, the states of the northeast extending south to include Virginia, West Virginia, and Kentucky, have traditionally been considered the territory of the National Building Code.

Standard Building Code 1997 -
Southern Building Code Congress International (SBCCI)

The Standard Building Code has been the predominant code for southeastern states including Louisiana, Arkansas Tennessee, North Carolina, South Carolina, Georgia, Florida, Alabama, and Mississippi.

2.3   State Authority

Model codes are developed at the national level, but they have no legal authority until individual states or local governments adopt them. Some states have a mix of national model codes; some states have none because there is no authority to adopt a code on a statewide basis. In the latter situations, local jurisdictions may elect to adopt the code of their choice and make local amendments. Some states have passed legislation that establishes the code that jurisdictions must adopt if they elect to adopt a code. This promotes uniformity and consistency across jurisdictional lines within the state and facilitates compliance and inspection.

Developers of new power generation technologies, whose products may be installed in non-traditional IPP locations (residential and small commercial facilities), will need to develop an industry standard for their product, lobby for its adoption/reference in the national model codes, and then work with state agencies to have the model code adopted and enforced. This is a long-term effort for which the DOE already has state outreach programs in place to provide assistance.

3.0   MANUFACTURING STANDARDS: "THE BOX"

As noted in Chapter 2, voluntary consensus standards are developed by a variety of organizationsjeanThese standards then may be referenced by model codes and adopted by state or local government organizationsjeanIEEE, ASME, ASTM, and NEMA are particularly active developers of consensus standardsjeanAlthough no manufacturing standards currently exist for microturbine generator sets as a complete, separate product, several standards from these organizations are relevant to the components of microturbines.

Underwriters Laboratories has worked with microturbine developers to list their product as an alternative to UL 2200, Engine-Driven Generator SetsjeanBeyond this, options currently available to the microturbine community include:

* request for revisions or additions to UL 2200, providing design specifications for microturbines

- development of a manufacturing or design standard specifically for microturbines by one of the standards-developing organizations mentioned earlier

- third-party testing and certification (see chapter 6), which may be acceptable to many code authorities.

The microturbine industry is not the first among DG developers to address code issues for technologically advanced onsite generation. The U.S. fuel cell industry has successfully developed a consensus standard, ANSI Z21.83, Fuel Cell Power Plants, which evaluates the construction safety and performance safety of a fuel cell power plant using natural gas or liquified petroleum (LP) gasjeanIn addition, fuel cell installations now are referenced in the National Fire Code.

PTC 50, Fuel Cell Power Systems Performance, contains methods and procedures for conducting and reporting fuel cell system testing, instrumentation to be used, testing techniques, and methods for calculating and reporting results.

3.1 ASME/ANSI B133, Gas Turbine Procurement

This suite of standards deal with procurement standards for gas turbines in industrial marine and stationary power applications. They are not currently considered applicable to microturbines.

3.2 UL 2200, Stationary Engine Generator Assemblies

This is the standard currently used by third-party testing organizations to list microturbines, although it is not an ASME- or IEEE-equivalent manufacturing or performance standard. It deals almost exclusively with electrical safety issuesjeanCurrently microturbines are not referenced by name; however, UL is reviewing their work performed for manufacturers to identify which sections will require revision to make allowance for microturbine systems.

As currently written, UL 2200 covers stationary engine generator assemblies rated 600 volts or less installed and operated in non-hazardous locations in accordance with NFPA-37, Standard for the Installation and Use of Stationary Combustion Engines and Gas Turbines; NFPA 99, Standard for Health Care Facilities; and NFPA 110, Standard for Emergency and Standby Power Systems.

Because UL 2200 is essentially a safety standard, it contains practical requirements for the design, construction and installation of any engine- or turbine-generator assemblyjeanNone of the basic performance data given for the microturbine designs being considered in this review is in direct conflict or in direct violation of this standard's requirementsjeanHowever, sections that address the prime mover (historically an IC engine) will require a comprehensive review for microturbine combustors, inlet temperatures and gas pressure, electrical output, etc. Please see chapter 4 for additional information on UL 2200 as a safety standard.

4.0 INSTALLATION AND OPERATION STANDARDS

Underwriters Laboratories (UL), the Electrical Generating Systems Association (EGSA), and the National Fire Protection Association (NFPA) have developed standards most relevant to microturbine installations. Those three standards are summarized in this chapter.

4.1 UL 2200, Stationary Engine Generator Assemblies

This standard deals almost exclusively with safety issues in relation to electrified equipment. It does not cover any mechanical performance standards of the prime mover and ancillary equipment. Currently microturbines are not referenced by name in the standard, although this is the central document by which third-party testing organizations are listing microturbine products as safe for operation.

The requirements spelled out in this standard cover stationary engine generator assemblies rated 600 volts or less that are intended for installation and use in ordinary locations in accordance with the National Electrical Code, NFPA-70; the Standard for the installation and Use of Stationary Combustion Engines and Gas Turbines, NFPA-37; the Standard for Health Care Facilities, NFPA-99; and the Standard for Emergency and Standby Power Systems, NFPA-110.

These requirements do not cover 1) generators for use in hazardous (classified) locations; 2) uninterruptible power source (UPS) equipment; or 3) generators for marine use.

The contents of UL 2200 are in terms of the following sections:

Construction of the Unit—Consists of essential features for general protection, controls, accessible circuits, and other mechanical and electrical subassemblies. Section 34 at the end of this part deals with requirements for "Protection of Service Personnel." This part of the standard mandates, among other requirements, a minimum thickness of the cast metal enclosures and sheet metal enclosures of the generator assemblies; a minimum thickness for nonmetallic enclosures (UL 746C); flammability limits (UL 94); and minimum thickness and maximum area requirements for glass-covered openings.

Mechanical Systems—Construction—Deals mainly with fuel systems and exhaust systems.

Mechanical Systems—Performance—Deals with several testing requirements, including tests for harmonic distortion, voltage and frequency fluctuation, salt sprays, grounding impedance, overcurrent protection calibration, and impacts, among others.

Marking—details and instruction manual

Manufacturing and Production Tests—Production-Line Dielectric Voltage-Withstand Test

Outdoor-Use Units—Requirements for these units supplement and, in some cases, modify the general requirements given above.

Standards for Components—Lists the UL standards with which components of the generator sets must comply.

Because UL 2200 is essentially a safety standard, it contains requirements primarily focused on electrical output safety performance measures that are mandatory for the design, construction and installation of any engine- or turbine-generator assembly. None of the basic performance data given for the microturbine designs being considered in this review is in direct conflict or in direct violation of this standard's requirements.

4.2 EGSA 101P-1995, Performance Standard for Engine Driven Generator Sets

The requirements defined in 101P were compared with specifications of a manufacturer's turbo alternators, either available or planned, with nominal outputs in the 45-kW to 200-kW range, and also with specifications for a 30-kW device.

The primary conclusion is that 101P is currently incomplete as a performance standard. It contains a mixture of qualitative, quasi-quantitative, and fully quantitative requirements together with tutorial sections that express no requirements. While it might be considered a baseline from which to develop a performance standard, it lacks sufficient detail to be useful. Finally, as 101P relies extensively on other referenced standards, it should be asked whether or not it represents redundancy.

Subsection 8.2 provides the greatest concentration of quantitative performance requirements within this standard. Of note, however, is the lack of cross-referencing all the other standards listed in Section 2. This appears to be an incomplete standard for determining the adequacy of a microturbine installation and operation.

4.2.1 Applications Criteria

Modes of Operation—Somewhat advisory in declaring the operating modes of the generator set must be considered when the purchaser negotiates requirements with the equipment supplier.

Generator Set Rating Definitions—Subsections 3.2.1 through 3.2.4 define the terms Emergency Standby Rating; Limited Running Time Rating; Prime Power Rating; and Industrial Rating. The Prime Power Rating requires a minimum momentary overload capability of 10%.

Application Classes—Subsections 3.3.1 through 3.3.4 define four application classes and their associated, progressively less severe, qualitative requirements on frequency, voltage, and waveform.

Criteria of Use—Subsections 3.4.1 and 3.4.2 distinguish land use from marine use.

Classification of Operation—Subsections 3.5.1 and 3.5.2 distinguish the character of Single Unit Operation from Parallel Operation.

4.2.2 Prime Mover

Engine Configuration—Requires assurance from the manufacturer that the engine is fit for service. Requires engine conformance with EGSA 101S and ISO 3046.

Engine Types—States focus of the standard is predominantly diesel but applies generally to gas turbine, gasoline, or natural gas engines.

4.2.3 Generators (Alternators)

Design Standards—Requires generators be designed to meet appropriate sections of NEMA MG1 and IEC 34. Note: The latter was not listed in Section 2. This subsection refers to maximum temperature rise defined in NEMA MG 1 and torsional compatibility of generator set components.

Ratings—Equipment rating stated in kW or kVA at assumed minimum power factor of 0.8 lagging and at specified voltage and current.

4.2.4 Voltage Regulators (See EGSA STD 100R)

Application—Defines when VR is needed.

Voltage Sensing—May be single or 3-phase.

Voltage Adjustment—±5% of nominal rated voltage

Voltage Regulation Accessories—Covers in 6.4.1 through 6.4.5 parallel operation, motor starting, under frequency protection, over voltage protection, and electromagnetic interference. Two of these are hard requirements (provision for reactive droop compensation and under frequency protection); the other three appear to be purely advisory.

4.2.5 Control and Monitoring Panel

Configuration—Refer to 101S.

Construction—Must conform with applicable NEMA and ANSI standards.

Instrumentation—Legibility recommendation for scales and metering accuracy requirements (±2% for ac electrical instruments and ±10% for engine instruments/indicators).

Current Rating—Current carrying capacity requirements for conductors and components (should be sufficient for service).

Identification of Components—General good human factors design requirements for component identification

Electrical Protection-Requirement for short-circuit protection and recommendations for other related practices, including some that belong in 7.5.

4.2.6 Complete Generator Set

Terms and Expressions—Defines quantities measured by representative voltage strip chart (8.1.1) and frequency strip chart (8.1.2). No requirements, just definitions.

Unit Starting and Load Acceptance—Contains three requirements: 1) emergency (must be started, on line and under load in 10 seconds or less); 2) standby (must be started, on line and under load in 1 minute or less); and 3) additional starting aids (use of aids; e.g., engine coolant and battery temperature control, glow plugs and/or oil heating) under severe conditions.

Rated Power—The generator set must be capable of producing its rated power at rated frequency, voltage, and electrical power factor, corrected to standard ambient conditions per SAE J1349.

Governor Performance for Application Classifications—Contains definitions and quantitative requirements relating to frequency regulation.

Voltage Regulation—Subsection 8.2.4.1 requires steady state VR no more than 2% between no load and full rated load (per EGSA 100R). It also defines requirements for random voltage variation and voltage sensitivity to temperature. Subsection 8.2.4.2 provides advice on the tailoring of self regulated alternator voltage.

Excitation Support System (Optional)—Defines 2.5 per unit exciter current requirement when applicable.

Unbalanced Loads—Recommends keeping unbalance to less than 20%; connecting no more than 1/3 of nominal kVA rating to any one phase of a 3-phase system and requirements for delta, delta-delta or open delta connections together with voltage unbalance limits.

Waveform and Telephone Influence Factor—Defines maximum allowable harmonic content and telephone influence factors for application classifications listed in Section 3.3.

Overspeed—The generator should tolerate 25% overspeed for 10 seconds without damage. Overspeed protection should activate at 20% over maximum rated synchronous speed.

Engine Cooling System—Defines cooling water temperature limit under specified conditions of antifreeze content, ambient temperature, elevation and radiator air-flow restriction.

Engine Air Cleaner—Defines type and performance requirements for engine air cleaners.

4.3 ASME B133.6, Gas Turbine Ratings and Performance

This standard was originally developed to provide a performance rating system for installation sites that do not have sufficient data on ambient conditions to use the manufacturer's ratings. Because it was last updated in 1994, it does not currently apply to the microturbine products commercially available today.

4.4 NFPA 37, Standard for the Installation and Use of Stationary Combustion Engines and Gas Turbines (1998)

NFPA 37 concerns the installation of engines for stationary power. NFPA 99 references NFPA 37 in the text and NFPA 110 references it in the appendix. Textual references make the standard mandatory. Appendix references are for information only. The International Mechanical Code (IMC) also references NFPA 37 for engines powering equipment and appliances.

Chapter 2 gives requirements for mounting, locating, and housing engines. Persons quoting generator sets requiring conformity with NFPA 37 should be aware of Par. 3-3.1(b) and (f). These often overlooked clauses requires lubricating oil over temperature shutdown (b) or indication (f) for engines over 100 horsepower.

Chapter 5 is the reference for liquid-fueled engine fuel tank and daytank installations. This chapter gives some details, such as a table for day tank steel thickness, but frequently references NFPA 30, Flammable and Combustible Liquids Code. Chapter 4 does the same thing for gas-fueled engines and references NFPA 54, National Fuel Gas Code and NFPA 58, Standard for Storage and Handling of Liquefied Petroleum Gases.

Chapter 6 covers exhaust piping and chimneys. It gives details for safe exhaust piping routing and installation. Examples are material of wrought iron or steel and drains for low points in the line. For chimneys, the standard references NFPA 211, Standard for Chimneys, Fireplaces, Vents and Solid Fuel Burning Appliances.

NFPA 37 has only seven pages of text but be especially aware of Chapter 10, Mandatory Referenced Publications. It lists some 13 additional standards that must be reviewed for each installation. 4.7

4.4.1 NFPA 37 Scope - Size Limits

The previous 7500-horsepower limitation in the scope of the standard has been removed. There is no minimal horsepower threshold for engines or gas turbines so microturbines would be subject to the requirements set forth in this standard.

This standard applies to fire safety for the installation and operation of stationary combustion engines and gas turbines. It also applies to portable engines that remain connected for use in the same location for a period of one week or more and that are used instead of or to supplement stationary engines. For engines used in essential electrical systems in health care facilities, also see NFPA 99, Standard for Health Care Facilities. For engines used in emergency power supplies, also see NFPA 110, Standard for Emergency and Standby Power Systems.

The term **combustion gas turbine** as used is inclusive of microturbines and, therefore, the standards requirements apply to microturbines. The standard does not address the electric generating component of the microturbine generator unit.

Chapter 3 of the standard addresses general requirements for engine locations (in, on, and outdoors near buildings), electrical installations in rooms containing engines (NFPA 70 by reference), engine wiring, and other general installation requirements (reference to applicable NFPA codes and standards), and to those portions of existing equipment and installations that are changed or modified establishes requirements for locating microturbine engines in, on, or near buildings.

Chapter 4 addresses gas piping and references NFPA 54 for systems at service pressures of 125 psig and less and NFPA 58 for LP-gas systems. It prescribes the minimal components of a gas train for engines as containing a manual shutoff valve, regulator, low-pressure switch, automatic safety shutoff valve, automatic control valve, manual leak test valve, and high-pressure manual reset switch with exceptions. Boosters or compressors, if used, shall be approved for the service intended and receivers, if used, must be stamped as complying with the ASME Boiler and Pressure Vessel Code.

Chapter 5 addresses Class 1 liquid fuels such as gasoline, gasohol, and alcohol, and liquid fuels other than Class 1 such as diesel fuel, fuel oils, jet fuel, and kerosene. The only microturbine concern in this chapter is a statement that LP-gas systems in the liquid phase must be installed per NFPA 58.

Chapter 6 presents general requirements related to lubricating systems and some requirements specific to gas turbine oil reservoirs. Technologies not employing lubricating oil reservoirs would have no interest in this chapter.

Chapter 7 addresses engine exhaust systems and clearances to combustible materials for exhaust gas temperatures less than 1400°F that would apply to microturbine installations.

Section 4 of Chapter 8 imposes control and instrumentation requirements on gas turbines. Each engine must be equipped with an automatic engine speed control, an automatic main speed control and overspeed shutdown control, a backup overspeed shutdown control that is independent from the main control, an automatic engine shutdown device for low-lubricating oil pressure (with exception), provisions for shutting down the engine from a remote location, provisions for shutting down, from a remote location, lubricating oil pumps not directly driven by the engine, an automatic engine shutdown device for high exhaust temperatures (with exception), and a means of automatically shutting off the fuel supply in the event of a flameout. The starting sequence must include a purge cycle.

Chapters 9 and 10 address operating and emergency instructions being readily accessible to personnel operating or maintaining the engine. Individuals responsible for the operation and maintenance of the engine should be familiar with the procedures.

4.5   NFPA 110, Emergency and Standby Power Systems

The emphasis of NFPA 110 is on reliability, performance, testing and maintenance. The NFPA Standards Council has ruled that NFPA 110 is primary to the NEC and NFPA 99 in these matters. NFPA 110 is much more specific with regard to generator-set installations than other codes.

NFPA developed the standard primarily for use by permitting authorities who needed a comprehensive guide beyond the various uncoordinated requirements in other codes. It addresses only the generator set and transfer switch. This standard was under contentious development for almost a decade, and any changes proposed for additional equipment can expect to be challenged.

The standard covers performance requirements for power systems (power sources, transfer equipment, controls, supervisory equipment, and all related electrical and mechanical auxiliary and accessory equipment) providing an alternate source of electrical power to loads in buildings and facilities in the event that the primary power source fails.

Section 2-3. Classifications of Emergency Power Supply Systems (EPSSs)

The standard classifies EPSSs as Types, Classes, Categories, and Levels.

Type indicates the maximum time in seconds allowed before the EPSS assumes the load. Thus, for a Type 10, the system must be fully operational within 10 seconds.

Class indicates the minimum time in hours that the EPSS will operate without refueling. Class 2 represents a system designed to operate for 2 hours.

Level defines the importance of the installation to life safety. Level 1 defines requirements for applications where failure could result in serious injury or loss of human life. Level 2 defines applications that are less critical to life. Level 3 refers to all others. No requirements for Level 3 are in the standard.

Levels 1, 2, and 3 are roughly equivalent to Emergency, Legally Required Standby, and Optional Standby in the NEC.

Unique to NFPA 110 is the requirement for prototype testing of the generator set. The supplier must show proof of performance under normal and adverse conditions before installation; this avoids problems otherwise not discovered until the installation startup, or later.

The standard requires about a dozen visual safety and shutdown indications at the generator set. It also calls for remote audible alarm for any of the conditions. It calls for prealarms where early attention might avoid a system shutdown.

During testing the system must perform all functions with results observed and recorded. Paragraph 5-13.2.5 calls for a two-hour full nameplate kW load test. The test need not be at rated power factor if the factory test was at rated power factor. Immediately after the load test and a five-minute cooldown, the system must demonstrate that it can pick up full kW load in one step.

4.6   ASME B133.8, Gas Turbine Installation Sound Emissions
This standard gives methods and procedures for specifying the sound emissions of gas turbine installations for industrial, pipeline, and utility applications. Included are practices for making field sound measurements and for reporting field data. This standard can be used by users and manufacturers to write specifications for procurement, and to determine compliance with specification after installation.

Some microturbine manufacturers have used this standard as a baseline for their product's noise profile. As it was originally written for larger gas trubine systems, however, the allowable thresholds outlined in this standard may be higher than those required by a municipal authority in a commercial or multi-unit residential setting.

5.0   INTERACTIONS BETWEEN MICROTURBINE GENERATOR-SETS AND OTHER BUILDING SYSTEMS, STRUCTURES, OR SAFETY ISSUES

Issues that may attract the attention of local building permit authorities include fire protection, egress, ventilation, electrical shock protection, and fuel supply. The principal codes that apply are outlined in this chapter. Figure 5.1 presents the most common installation requirements a microturbine generator set may be subject to in a commercial setting.

5.1   NFPA 70-99, National Electrical Code
The purpose of the National Electrical Code (NEC) is the practical safeguarding of persons and property from hazards arising from the use of electricity. It covers installations of electric conductors and equipment within or on public and private buildings or other structures and conductors and equipment that connect to the supply of electricity. Historically, it has not controlled installations under the exclusive control of electric utilities for the purpose of communications, metering, generation, control, transformation, transmission, or distribution of electric energy. Such installations shall be located in buildings used exclusively by utilities for such purposes, outdoor on property owned or leased by the utility, on or along public highways, streets, roads, or outdoors on private property by established rights such as easements.

Many NEC requirements, important to on-site power, lie buried in articles not obvious by their titles to have relationship to generators. The

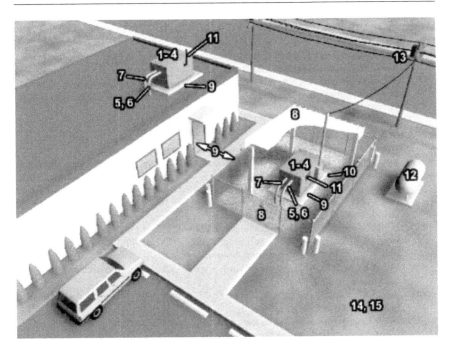

Figure 5.1. Typical Installation Requirements for Microturbine Generators in Commercial Settings

Legend

1. Component or OEM manufacturing and system integration: relevant standards from ANSI, ASHRAE, ASME, ASTM, EGSA, IEEE, NEMA, UL
2. NFPA 37, sections 8-4.1-2, combustion gas turbines must be installed with (1) an automatic main speed control and overspeed shutdown control, (2) a backup overspeed shutdown control independent from #1, (3) automatic engine shutdown device for low lubricating oil-pressure (exempt when unit is constantly attended), (4) remote engine shutdown capability, (5) remote lubricating oil pump shutdown capability, (6) remote shutdown capability for high exhaust temperatures, and (7) automatic fuel shutoff during a flameout. Additionally, the turbine starting sequence "shall include a purge cycle that produces a nonflammable atmosphere in the turbine and exhaust system prior to introduction of the fuel." (section 8-4.2)
3. NFPA 37, section 4-4.3, unattended engines shall have both a zero governor-type regulator installed AND an auxiliary shutoff valve installed ahead of any flexible connector to other controls.
4. NFPA 37, sections 6-2.1-4, requirements for gas turbines utilizing an oil lubrication system.
5. All fuel gas systems utilizing service pressures under 125 psig must be installed and operated in accordance with NFPA 54, the National Fuel Gas Code. Fuel piping must be steel or other metal, and in compliance with NFPA 30, Flammable and

(Continued)

Figure 5.1. Legend (*Continued*)

Combustible Liquids Code. Piping protection is outlined in MSS SP-69, Pipe Hangers & Supports - Selection and Application.

6. All pressure-boosting equipment must be certified (by stamp) for design, construction and testing according to ASME Boiler and Pressure Vessel Code, Section VIII, Division I.

7. NFPA 37, section 3-5.3.1-3, all wiring and batteries must be protected from arcing and shorting. Note: all wires, with the exception of ignition or microprocessor wiring and thermocouples, must be stranded annealed copper.

8. Enclosure - although NFPA 37 does not specifically address enclosures, the local building inspector will probably require (1) the cabinet to meet NEMA standards for explosion-proof enclosures, and (2) the integrated unit (prime mover, generator, pipes and wires and all controls & valves) to be protected from natural elements (wind, sun, precipitation, detritus) and vehicular impact.

9. NFPA 37, section 3-1.4.1 (outdoor installations): unit must be a minimum of 5 ft. from any combustible walls and bldg. openings. Also NFPA 37 section 3-2, foundation to be made of non-combustible materials.

10. NFPA 37, section 3-1.3.1, same minimum space (5 ft.) and foundation requirements as outdoor installations.

11. ASHRAE Handbook—Fundamentals (1993) is recommended reference for design of exhaust air discharge system.

12. All LP gas systems (liquid or vapor phase) must be installed in accordance with NFPA 58, Standard for the Storage and Handling of Liquid Petroleum Gases. The Flammable and Combustible Liquids Code, API (American Petroleum Institute) 620, Design and Construction of Large Welded Low-pressure Storage Tanks, may also apply.

13. Grid Interconnection (standard currently under development by IEEE SC 21, Richard DeBlasio, chair)

14. Local zoning ordinances (definition of hazardous materials and relation to residential zones, distance to property line and rights-of-way, access by local fire and safety authorities, etc) may need to be consulted in some areas.

15. Fire Protection - Local building inspectors will require a fire risk evaluation be performed for each installation with respect to design, layout, and operating conditions of the unit. From that analysis the inspector may require any or several of a variety of fire protection systems (portable vs. fixed systems, foam or gaseous extinguishers, automatic sprinklers or dry chemical fire suppression systems).

NEC does not have an article on portable power sources, yet requirements exist in Articles 210-Branch Circuits, 230-Services, 240-Overcurrent Protection, 250-Grounding and 305-Temporary Wiring. Article 517 has important requirements for emergency systems in health care facilities.

Because on-site power must be compatible with utility-furnished power, many of the requirements of other articles also apply to on-site power. For example, Article 250-5 tells what systems must be grounded. These requirements apply, with certain listed exceptions, regardless of the ownership of the power-generation unit.

5.1.1 Article 700, Emergency Systems A

Article 700 has five parts. Most important to the emergency power source are Parts A, General, and C, Sources of Power. The following items are of particular importance to onsite generators:

- Article 700 covers emergency systems that are legally required and classed as emergency by the governmental agency having jurisdiction or by other codes (hospitals, public facilities, missioncritical infrastructure such as air traffic control towers, etc.)

- Tests at maximum anticipated load must be witnessed, performed periodically, and a written record maintained.

- Other loads (including peak shaving) can be served if an automatic load management system ensures adequate emergency power.

- Automatic transfer switches must prevent inadvertent interconnection of normal and emergency sources (but can be bypassed).

- Audible and visual signals must indicate (1) any disturbance of the emergency power source, (2) that the battery is carrying load, (3) that the charger is not functioning, or (4) that a ground fault exists. 5.4

- In the event of an emergency, current must recommence within 10 seconds and can be from storage battery, generator set, uninterruptible power supply (UPS) system, a separate service, or individual equipment for emergency illumination.

- A generator set driven by a prime mover must have automatic or "black" start capability with a minimum of 2 hours full-load fuel supply on-site, unless there is a low probability of simultaneous failure of both the fuel delivery system and the electric utility.

- A starting battery must have an automatic charger independent of the generator set. A 15-minute time delay is required before shutdown of the set.

5.1.2 Article 701, Legally Required Standby Systems

Article 701 refers to applications less critical to life, such as heating and refrigeration systems, communications systems, and industrial processes. The NEC itself does not require standby systems, but outlines the requirements should some other authority specify one. This generally includes illumination and/or power. The requirements are somewhat less stringent than those for Article 700.

A major difference between Article 700 and 701 is that 701 does not require standby circuits to be in separate raceways. It does not require any indication for ground faults, and it allows 60 seconds to start the standby power system, rather than the 10 seconds required by 700-2.

5.1.3 Article 702, Optional Standby Systems
Article 702 briefly addresses standby systems intended to protect private property, where life safety does not depend upon the performance of the system. Typical installations include industrial and commercial buildings, farms and residences. The note specifically mentions heating and refrigeration, data processing and communication systems, and industrial processes. Only permanently installed equipment applies; it does not apply to portable generator sets.

The equipment is required to simply have sufficient capacity to supply the loads intended for simultaneous operation. It does require a transfer switch. Signals are required, where practicable, for showing derangement of the source and the successful assumption of the load. Article 702 does not require any testing or keeping of records at installation or periodically afterward.

5.1.4 Article 705, Interconnected Electric Power Production Sources
Article 705 covers installation of one or more electric power production sources in parallel with a primary source of electricity such as a utility or on-site electric power source. The code addresses some issues related to grid-connection but these requirements were not developed around microturbine technology and cogeneration applications. Efforts to establish uniform and consistent requirements for protective features specific to microturbine and other DG technologies that electric utilities would accept should focus on this code.

705-12, Point of Connection Requires interconnection at the service disconnecting means. Two exceptions apply. One, if the system qualifies as an integrated electric system and complies with all provisions of Article 685. Two, if the system non-utility sources total more than 100kW or more than 1000 volts. The second exception also requires assured maintenance and supervision and protective safeguards.

705-22, Disconnect Device Requires an accessible disconnecting means for each power source. (Utilities may require that this device be accessible to linemen and may require that the on-off position of the device be visibly verifiable.)

705-40, Loss of Primary Source Requires that upon loss of the primary power source the on-site source automatically disconnect from the primary source. It shall also not reconnect until the primary source returns.

5.1.5 Article 445, Generators

Article 445 has few if any requirements that most generator manufacturers do not meet with a standard product. It has the usual requirements for enclosure, protection of live parts and nameplate markings. It does require over current protection but allows a wide choice of methods including inherent protection.

5.1.6 Other Articles
305-6(a) Ground Fault Circuit-Interrupters (GFCI) Temporary wiring requires GFCI for receptacles. The exception, however, exempts portable or vehicle-mounted generators not more than 5 kW, provided both lines are insulated from the frame and all rounded surfaces. The system must have an equipment ground. Some authorities, including some generator set manufacturers, question the safety of an ungrounded system, however this section permits it.

250-5(d), Separately Derived System. This section requires grounding a separately derived system meeting the requirements of 250-5(b). A generator may or may not be a separately derived system.

A fine-print note explains that an on-site generator is not separately derived if the neutral is solidly interconnected to a service-system supplied neutral. Such generators should not be separately grounded. Other generators are separately derived and must be grounded if they can be grounded at 150 volts or less to ground or supply 480/277 volts.

517-65(b). Requires the alternate source of power to be a generator set located on the premises. In this case the NEC does not depend on any other code to require an emergency system and a generator set. Similar requirements are in other sections of Article 517 for other health care facilities. Strictly speaking, Articles 480-Storage Batteries and 690-Solar Photovoltaic Systems also concern on-site power sources. We will not try to cover those articles. Those systems are outside the scope of this publication.

5.2   NFPA 101, The Life Safety Code

This code deals with life safety from fire and similar emergencies. It addresses construction, protection, and occupancy features necessary to minimize danger to life from fire, including smoke, fumes, or panic. A minimum criterion for the design of egress facilities to permit prompt escape from buildings has been established, along with other considerations related to life safety.

Chapter 5, Means of Egress

Chapter 5 of the Life Safety Code is the chapter most pertinent to microturbine product developers. Section 5-9 outlines the general requirements for emergency lighting; 5-9.2.1 requires emergency illumination for a period of 1.5 hours. Illumination must be an average of 1 foot-candle; illumination may decline to 0.6 and 0.06 at the end of the 1.5 hours.

The section refers to NFPA 110 for installation, testing, and maintenance of the generator set. The standard mentions only emergency lighting and exit signs. (It refers to other standards for fire detection equipment, fire alarms, elevators, and escalators. Those standards frequently will call for fire detection, alarms, and one elevator to be on the system.)

Chapters 8-30 tell whether emergency lighting is required for each specific occupancy. For example, in Chapter 10, New Educational Occupancies, Section 10-2.9 requires emergency lighting in all interior stairs, corridors, and normally occupied spaces.

In general, schools, most health care facilities, places of assembly, correctional institutions, hotels, dormitories, apartment buildings, and certain mercantile buildings require emergency lighting of means of egress.

5.3   Fuel Gas Codes

Of all the safety codes, these are some of the most necessary. They rank with the NEC in their importance in saving lives.

5.3.1 International Fuel Gas Code (IFGC) 1997

This code was developed to supplant national fuel gas codes and is recognized by the three regional building code (BOCA, ICBO, SBCCI) organizations in the United States. As of January 1, 2000, this will replace the U.S. National Fuel Gas Code (NFPA 54, from ANSI Z223.1).

The IFGC applies to the installation of fuel gas piping systems (design, materials, components, fabrication, assembly, installation, testing, inspection, operation, and maintenance), equipment fueled by gas (installation and combustion), and "related accessories" (ventilation air and venting) for

- systems using gas at an operating pressure of 125 psi or less (covering all microturbines)

- LP gas piping systems operating at 20 psig (gauge) or less

- flammable gas-air mixtures operating at 10 psig.

The code specifically does NOT cover:
- portable LP gas equipment (unless connected to a fixed piping system)

- farm equipment

- raw material (feedstock) applications

- oxygen-fueled cutting and welding systems

- industrial acetylene and acetylenic compounds, hydrogen, ammonia, carbon monoxide, oxygen, and nitrogen

- industrial fuel processing (petroleum refineries, pipeline compressor/pumping stations, refinery tanks, natural gas processing plants)

- chemical plants where flammable or combustible liquids/gases are used in or created as a byproduct of a chemical reaction

- liquid natural gas (LNG) installations

- power plants

- "proprietary" (utility-owned) items

- temporary LP gas systems at a construction site

- vehicular LP and natural gas systems

- systems operated by gas LDC

- general building design and construction.

It is vitally important to note that Section 108.7.2 gives code officials the authority to disconnect any energy source supplied to a building (or other structure regulated by this code) that is deemed "hazardous or unsafe." Written notice gives the building owner 24 hours to disconnect, and, in the case of immediate "danger to life or property," the code official can immediately disconnect without prior notice.

Section 304, Combustion, Ventilation and Dilution Air Applies when 1) gas-fueled equipment is installed inside a building, and 2) the equipment requires air for combustion, ventilation or dilution of flue gasses from within the building. Direct venting to the outside overrides this section.

Chapter 4, Gas Piping Installations Chapter 4 establishes criteria for the minimum/maximum pipeline sizes for a given appliance, construction materials, shut-off valve locations, metering devices, etc. There is nothing in this section that specifically prohibits the installation of a microturbine unit.

Chapter 5, Chimneys and Vents This section covers the installation, maintenance, repair, and approval of factory-built chimneys, chimney liners, vents, connectors, and masonry chimneys. Every fuel-burning appliance is required to discharge all combustion products to the outdoors. Section 501.8 lists exempted equipment, of which microturbine units do not qualify (e.g., clothes dryers, cooking ranges, refrigerators, counter appliances, etc.)

Microturbines, if subjected to this section, will be declared a category III or IV appliance. In residential applications, the venting system would default to the manufacturers' listed equipment. In commercial or industrial applications, a chimney, ventilating hood, and exhaust system all would be required.

Chapter 6, Specific Appliances

Section 615.1, Engine and Gas Turbine-Powered Equipment, requires that permanently installed equipment powered by internal combustion engines and turbines be installed in accordance with the manufacturer's installation instructions and in accordance with NFPA 37 (Stationary Combustion Engines and Gas Turbines).

5.3.2 NFPA 54, National Fuel Gas Code

This is the governing code used by many local gas utilities and officials of federal, state, and local governments to judge the acceptability of fuel-gas installations. Many appliance manufacturers as part of their certified installation instructions also reference the code.

5.3.3 NFPA 30, Flammable and Combustible
Liquids Code, and NFPA 58, LP-Gas Code

These two standards address the storage and handling of fuel liquids and, as such, do not impact the first-generation microturbine products.

5.3.4 International Mechanical Code - 2000

The International Mechanical Code regulates and controls the design, construction, quality of materials, erection, installation, alteration, repair, location, relocation, replacement, addition to, use, or maintenance of mechanical systems that are permanently installed and utilized to provide control of environmental conditions and related processes within buildings. The owner or agent shall be responsible for maintenance of mechanical systems. To determine compliance with this provision, the code official shall have the authority to require a mechanical system to be reinspected. Code officials should be educated on microturbine application and operation to minimize re-inspections.

Section 920 requires that gas turbines, including fuel storage and piping, meet the requirements of NFPA 37. Permanently installed equipment powered by the turbine must be installed in accordance with the manufacturers installation instructions and in accordance with NFPA 37.

Chapter 12 provides requirements for hydronic piping systems that are a part of a heating system and requires that potable hot water systems meet the plumbing code. Certain CHP installations may involve these requirements. Chapter 13 governs the installation, modification, and maintenance of fuel-gas piping systems. The applicability of this code to piping systems extends from the point of delivery to the connections with each utilization device and includes the design, materials, components, fabrication, assembly, installation, testing, inspection, operation, and maintenance of such piping systems. NFPA 54 is referenced for testing, inspecting, and purging gas piping systems.

6.0   THIRD-PARTY TESTING AND CERTIFICATION

Until a national consensus standard is available for microturbines, developers can employ third-party testing organizations to certify the safety and operability of their product. Test results and design specifications are compared against a suite of existing codes and standards related to the product.

The drawback to this method is twofold: 1) without a national standard referenced in the model codes, every local code jurisdiction can require a battery of additional tests and design reviews, or may not permit the unit at all; and 2) education of code officials lasts only as long as that official in that office.

This chapter provides brief outlines of the most common testing and certification organizations in the United States.

6.1   CSA International

CSA International (formerly the American Gas Association Certification Labs) provides product certification, management systems registration, and information products. The organization is a Nationally Recognized Testing Laboratory (NRTL) by the U.S. Occupational Safety and Health Administration (OSHA). Certification is recognized by federal, state, provincial, and local authorities.

Certification options include the following:

Category Certification—Manufacturer arranges the tests, at own facility or at another approved facility. Manufacturer documents the results and determines whether the product meets the necessary requirements.

Shared Certification—Manufacturer performs the testing and prepares the Certification Report, which is then reviewed by CSA. CSA examines the product, reviews and validates the data, and may spot-check the testing process before granting the appropriate certification.

Witness Testing—Products are tested at manufacturer's facility by own staff, but in the presence of a CSA International representative. Based on the results of the testing, the CSA International representative then prepares the Certification Report and issues the Certification Letter.

Model Certification—Manufacturer sends product samples to CSA International for testing at its laboratory. CSA International then writes the Certification Report and, if the product qualifies, issues the certification.

6.2 Intertek Testing Services

Intertek Testing Services Scania, formerly known as the Electrical Testing Laboratory, provides independent third-party commodity sampling, inspection, testing,and related marine surveying services to the petrochemical, fertilizer, coal, mineral, agricultural, and related industries.

ITS Quality Systems tests electrical and electronic goods from toys to telephones to appliances and medical equipment for product safety, electromagnetic compatibility, and both absolute and relative product performance. ITS also inspects and tests textiles, clothing, carpets, and consumer. The lab also provides assessment and registration services to the ISO-9000, QS-9000 (automotive), and the ISO-14000 (environmental) standards.

6.3 Hartford Steam Boiler

The Hartford Steam Boiler Inspection and Insurance Company (HSB) is the world's oldest (circa 1866) and largest insurer of power, process, and production equipment. The organization provides the following services:

ASME Code Inspection Services—An Authorized Inspection Agency (AIA) accredited by the American Society of Mechanical Engineers (ASME) for the inspection of boilers, pressure vessels, nuclear components and process and power plants. Services also include Pressure

Vessel and Piping Design, Finite Element Analysis, and Quality Assurance Program Development.

Department of Transportation Services—HSB offers assistance in complying with U.S. Department of Transportation and Canadian Transport Commission regulations for tank trucks and other devices/cylinders used in over-the-road and rail transportation.

International Codes and Standards Services—Both local and international codes and standards consulting, including inspection and technical training services.

Third-Party Inspection Services—HSB's Technical Resource Support Group ensures that the manufacture and/or installation of equipment by vendors meets the appropriate standards and specifications.

Pressurized Equipment Testing—HSB's Pressure Equipment Technologies provide inspection surveys, testing, and condition evaluation of critical powerhouse equipment including boilers, deaerators, and pressure vessels. The surveys evaluate the physical conditions of the equipment and provide a detailed report with outlined recommendations for corrective actions and life-expectancy projections.

6.4   Underwriters Laboratories Inc.

Services available from Underwriters Laboratories Inc. are detailed in Chapter 2 of this report.

6.5   National Evaluation Service

The National Evaluation Service, Inc. (NES) is an independent, not-for-profit organization that conducts a voluntary program of evaluation for building materials, products and systems. NES produces technical reports that describe a building construction material or product, together with a list of conditions necessary for compliance with each of the model codes, as promulgated by Building Officials and Code Administrators International, Inc. (BOCA), International Conference of Building Officials (ICBO), and Southern Building Code Congress International, Inc. (SBCCI) as well as the International Codes of the International Code Council (ICC).

The National Evaluation Service is a member of the World Federation of Technical Assessment Organizations (WFTAO), founded to accelerate the dissemination of technical information regarding performance testing and certification of new technologies. The organization's members represent 14 countries, including Australia, Brazil, Canada, Denmark, France, Hungary, Israel, Italy, Japan, New Zealand, South Africa, Spain, the United Kingdom and the United States.

7.0 OTHER BARRIERS TO MICROTURBINE PRODUCT ACCEPTANCE

Potential barriers to the acceptance of microturbine products exist in current regulations, codes, and standards in four primary areas: emissions control, grid interconnectivity, zoning, and noise control.

7.1 Emissions Control
Regulations for the control of emissions are in place at federal, state, and local levels.

7.1.1 Federal Regulations
The Clean Air Act (1963) and its amendments (1965, 1967, 1970, 1977, and 1990) provide the critical statute for restrictions on electric generation technologies. Title I establishes national ambient air quality standards (NAAQS) that prescribe the maximum permissible concentration of pollutants allowed in ambient air. Specifically, the Act requires the U.S. Environmental Protection Agency (EPA) to establish standards for six criteria pollutants:

- carbon monoxide (CO)
- nitrogen oxides (NO_x)
- sulfur dioxide (SO_2)
- particulate matter (PM)
- ozone
- lead

In general, the responsibility for reducing air pollution levels has been assigned to the states. Each state is required to develop a State Implementation Plan (SIP) providing for the implementation, maintenance, and enforcement of measures to attain the ambient air standards by the deadlines prescribed by the Clean Air Authority (CAA). The EPA has oversight authority for each state's SIP and may direct a state to revise its SIP if necessary.

Two elements that a SIP must contain are federal New Source Performance Standards (NSPS) and New Source Review (NSR) rules. The NSPS specify maximum pollutant emission rates for various processes, including combustion equipment. The EPA has assigned such performance standards for SO_2, NOX, and PM. NSPS are based on the level of control that can be achieved by the best demonstrated technology. NSR rules govern the permitting of new emissions sources and are triggered if a new source emits or has the potential to emit at an annual rate specified by the NSPS.

NSR rules distinguish between attainment and nonattainment areas with less stringent Prevention of Significant Deterioration (PSD) rules applying to attainment areas. The trigger for PSD rules is 250 tons/year for any regulated pollutant. Nonattainment areas are differentiated in classes based on severity of ambient pollutant concentrations: marginal, moderate, serious, extreme, and severe. NSR thresholds for nonattainment areas are indicated in Table 7.1.

To construct and operate a new power plant or DG facility (or to make major modifications to an existing plant) within a nonattainment area, the owner needs to obtain a permit from the state environmental agency if NSR levels are exceeded. The NSR process requires the owner to analyze alternative locations, sizes, production processes, and control techniques, and to demonstrate that the plant benefits outweigh its environmental and social costs. Facilities also are required to have control technology that meets the standard for lowest achievable emission rate (LAER). The control technology required to meet LAER is established by each state on a case-by-case basis for each emission source as it is permitted.

Furthermore, the owner of the plant is required to purchase offsets for each criterion pollutant that is in nonattainment. The EPA requires that emission offsets provide a positive air quality benefit to the area. Owners are therefore required to obtain more than one offset for each unit of pollutant emitted. The offset ratio depends upon the extent to which the region is in nonattainment. This offset requirement has promoted the establishment and trading of emission reduction credits for NO_X and volatile organic compounds (VOCs) among industries in 12 states.

The process for reviewing new facilities is slightly different in attainment areas. In these areas, owners also are required to obtain a permit to construct and operate new plants (or to make major modifications to existing plants), to ensure that new pollution sources do not make the region slip into nonattainment. These PSD permits require a review of the air quality impacts of the proposed facility.

Table 7.1. NSR Thresholds for Nonattainment Areas

Pollutant	Area Designation	Threshold (Tons per Year)
Ozone precursors (NO_X, VOC)	Marginal Serious Severe	100 50 25
Inhalable Particulate Matter (Pm10) and Pm10 Precursors (NO_X, SO_2, VOC)	Moderate	100
Carbon Monoxide	Serious	70
Nitrogen Oxides, Sulfur Dioxide	Any nonattainment area	100

New plants are required to install best available control technology (BACT) for all pollutants regulated under the CAA. The control technology required to meet BACT is established by each state on a case-by-case basis for each emission source.

Historically, less stringent controls apply to existing electric utility facilities in attainment areas due to grandfathering statutes. This provides clear advantages when competing with new sources that require specific emission controls as specified by NSPS (Biewald et al. 1998). This situation may be changing as a host of new regulatory initiatives could result in more stringent controls for existing facilities, especially coal-fired facilities. These controls could have a significant influence on the cost of power from coal-fired facilities, making them less attractive in a competitive marketplace.

A DG project, particularly one based upon hydrocarbon combustion such as a microturbine, could be impacted by a suite of other federal laws, including

- Clean Water Act (1987)
- Resource, Conservation and Recovery Act (1976)
- Occupational Safety and Health Act (1970)
- Toxic Substances Control Act (1976)
- Endangered Species Act (1973)
- Coastal Zone Management Act (1972)
- Historic Sites Act.

7.1.2 State Regulations

Each state is responsible for implementing programs that conform to the mandates of the federal CAA and associated amendments. Progressive states with areas experiencing severe air quality problems (e.g., Los Angeles) have implemented even more stringent air quality standards than current federal mandates. Regardless of the criteria, the responsibility for implementation and enforcement falls on air pollution control officials at the local level.

The potential for diverse state requirements can lead to inconsistent requirements that pose barriers or opportunities for microturbines. Differing requirements could allow more centralized power generation technologies (combined-cycle gas turbines) to choose which state had the least stringent requirements, while the power could be transmitted to the demand location. Conversely, inconsistent or uncertain requirements might be an added incentive for construction of distributed generation capacity that has low air contaminant signatures (renewables, fuel cells) not subject to state permitting regulations.

7.1.3 Local Environmental Regulations

Based on the projected emission signatures of microturbines, it is anticipated that federal regulatory requirements for NSR will not be triggered; local agencies will therefore be the primary regulatory authority overseeing microturbine installations relative to environmental concerns. Depending on the attainment status of the local area, permitting thresholds specified by Air Pollution Control Districts (APCDs) may limit the operating schedule or emission limits of a microturbine if a predetermined power rating is exceeded. In addition, control technologies may be required. If the permit thresholds are not triggered, microturbines will be exempt from local air quality regulations. It is likely that permitting requirements will change over the next several years. However, these changes may broaden the regulatory envelope to include microturbines.

7.1.4 Emission Signature of Microturbines

Of the different air and liquid (e.g., contaminated cooling water, oil seepage) emissions associated with microturbines, air emissions will have the greatest influence on a project's permitting. NO_X and CO_2 emissions are the critical emission categories, given their magnitude for conventional electric generation technologies. Although SO_2 also is an important emission category for traditional electric utilities, SO_2 emissions are expected to be negligible in microturbines.

The air emission signatures of selected DG technologies are shown in Table 7.2. These values are based on theoretical calculations or laboratory source testing (Cler and Lenssen 1997; NREL 1995). The actual emission signature for each technology is design-specific and dependent upon various locational criteria.

Fuel cells, with their electrochemical process of producing electricity, have the cleanest emissions profile, followed by microturbines and reciprocating engines, respectively. With the exception of CO_2, microturbines also exhibit generally low emissions for all classes of regulated pollutants.

Table 7.2. Emission Profiles of Selected DG Technologies

Distributed Generation Technology	Pollutant, lb/MM Btu			
	NO_X	CO_2	CO	SO_2
Microturbine	0.4 - 0.21	119	0.11	0.0006
Internal Combustion Engine (Gas)	3.1	110	0.79	0.015
Internal Combustion Engine (Diesel)	2.8	150	1.5	0.3
Fuel Cell	0.003	1	—	0.0204

Source: NREL (1995); Cler and Lenssen (1997)

7.2   Grid Interconnectivity

The most contentious and expensive portion of any microturbine installation is the unit's relationship with the local electrical distribution system and the end-user's current electric service provider. Most microturbine installations are not currently designed to provide 100% of peak load, and the owner/operator of the local electrical distribution company, or LDC, is expected to allow the microturbine to connect for grid-parallel operation and to provide backup service, should the unit fail. (In the vast majority of customer cases today, the LDC still owns the generation capacity serving that customer, as well).

The prospect of introducing thousands—or hundreds of thousands—of independently operated electricity-generating units onto distribution feeders gives rise to a variety of safety, system stability, and competitive issues. IEEE SCC21 has convened under Richard DeBlasio of NREL to develop P1547, "Standard for Distributed Resources Interconnected with Electric Power Systems." The consensus standard is expected to be completed by 2002 and will provide performance requirements for safety and system (grid) stability, including

- prevention of out-of-range voltage, frequency, harmonics and power factor
- disconnection for faults
- prevention of transient magnification
- limitation of DC injection
- limitation of voltage flicker induced by the generator
- prevention of disruptions on the utility system
- provisions for isolation
- integration with existing grounding system
- synchronization
- control of directional power flow
- prevention of islanding beyond point of common coupling
- prevention of re-energization beyond the point of common coupling
- detection of loss of grid: voltage or frequency disturbances 7.7

- single-phase fault detection for three-phase interconnection
- feeder reclosing coordination considerations
- regulators: power system and distributed generator voltage regulators, potential overvoltage conditions of islanded operation
- harmonics
- combination of harmonics and current imbalance
- voltage flicker
- islanding
- characteristics and properties of distribution transformers
- technology documentation
- electric power system documentation.

7.3   Zoning Ordinances

Zoning ordinances were developed to provide minimum standards for protection against incompatible uses on adjacent properties. In addition, zoning ordinances promote the health, safety, and general welfare of the inhabitants of a given municipality, or protect and conserve the value of the property within that city. For this purpose, the height, number of stories, size of buildings and structures, size and width of lots, the percentage of lot that may be occupied, the size of yards, courts and other open spaces, parking structures, the density of population and the location and use of buildings, structures and land for trade, marine business, industry, agriculture, residence, or other purposes may be regulated.

Although zoning ordinances generally do not address either standby or continuous-duty powergeneration systems in residential or commercial districts, it is possible that the advent of widespread DG systems in residential zones will require a review of the zoning literature, particularly what constitutes a "prohibited use" ("any use which is injurious, noxious or offensive by reason of the emission of odor, fumes, dust, smoke, vibration, noise, lighting or other cause"). Some local zoning ordinances are much more intimately involved in local building codes, including egress and fire protection measures. For this reason, it is possible that municipal planning departments will perceive the DG unit as a substantial alteration or addition to the existing use of a building, and require a zoning review as part of the permitting process.

The Building Codes Assistance Project (BCAP) has created an online forum for discussion of residential and commercial model energy codes. The purpose is to facilitate exchanges of information relating to energy codes. To subscribe, send an email to: bcap_energycodes_forum@ase.org. with the word SUBSCRIBE in the body of your message.

7.4   Noise Level

Noise level is a municipal issue and varies by jurisdiction. In general, this should not present a hurdle for microturbines. As an example, Capstone's Model 330 is rated at 65 dBA at 10 meters, while many outdoor sections of air conditioners and heat pumps are rated at 78 to 82 dBA.

8.0   CONCLUSIONS AND RECOMMENDATIONS

The review of standards and codes documented in this report forms the basis for the conclusions and recommendations presented in this chapter.

It appears that microturbine generator sets can comply with the existing model codes for installation and operation, despite the lack of any reference to their technology by name. However, as detailed in Chapter 2, as additional microturbine products enter the marketplace it will become increasingly important that they are referenced—by name—in the model codes, a process that must begin with a manufacturing or system integration standard. Several vendors have already approached Underwriters' Laboratories for certification of their product under UL 2200. This will give the industry certification for electrical safety and ease the local permitting process in many jurisdictions. But UL 2200 is a standard, not a model code. The National Electrical Code, Mechanical Code, International Fuel Gas Code, Life Safety Code, and International Building Code must all eventually be modified to reference, if not microturbines, then onsite power generation systems that include microturbines.

Over the past several years the fuel cell community has been successful in developing performance and installation standards that will be referenced in the year 2000 editions of the model codes. This effort was coordinated by the U.S. DOE, and can be used as a roadmap for other distributed generation systems.

Of equal importance, the entire industry of advanced, onsite power generation technologies need performance and reporting standards that can be referenced by federal, state and local authorities for operation permits, emissions evaluation and credits and energy-efficiency incentives. To this end, the U.S. Department of Energy should investigate support for a microturbine-specific performance standard, and for those applications most relevant to the emerging DER market—combined heat & power, premium power quality.

And despite the absence of any referenced code barriers to microturbines, a general lack of education among code officials on the differences between DER installations and traditional backup generators, is in fact hampering their deployment. A general State education program, providing code officials with hands-on experience and an opportunity to explore DER technologies before they receive a permit application, would support all stakeholders in the U.S. DER industry.

Appendix A
Selected IEEE Standards Relevant to Onsite Power Generation
421.1-1986 (R1996) "IEEE Standard Definitions for Excitation Systems for Synchronous Machines"
421.2-1990 "IEEE Guide for Identification, Testing, and Evaluation of the Dynamic Performance of Excitation Control Systems"
421.4-1990 "IEEE Guide for the Preparation of Excitation System Specifications"
505-1977 (R1996) "IEEE Standard Nomenclature for Generating Station Electric Power Systems"
665-1995 "IEEE Guide for Generating Station Grounding"
666-1991 (R1996) "IEEE Design Guide for Electric Power Service Systems for Generating Stations"
803.1-1992 "IEEE Recommended Practice for Unique Identification in Power Plants and Related Facilities—Component Function Identifiers"
928-1986 (R1991) "IEEE Recommended Criteria for Terrestrial Photovoltaic Power Systems"
929-1988 (R1991) "IEEE Recommended Practice for Utility Interface of Residential and Intermediate Photovoltaic (PV) Systems"
P929, D10 Feb 1999 "Draft Recommended Practice for Utility Interface of

Photovoltaic (PV) Systems"

1046-1991 (R1996) "IEEE Application Guide for Distributed Digital Control and Monitoring for Power Plants"

1145-1999 "IEEE Recommended Practice for Installation and Maintenance of Nickel-Cadmium Batteries for Photovoltaic (PV) Systems"

1150-1991 (R1998) "IEEE Recommended Practice for Integrating Power Plant"

1159-1995 "IEEE Recommended Practice for Monitoring Electric Power Quality"

1262-1995 "IEEE Recommended Practice for Qualification of Photovoltaic (PV) Modules"

1346-1998 "IEEE Recommended Practice for Evaluating Electric Power System Compatibility with Electronic Process Equipment"

C37.90.1 "IEEE Standard Surge Withstand Capability (SWC) Tests for Protective Relays and Relay Systems"

C37.95 "IEEE Guide for Protective Relaying of Utility-Consumer Interconnections"

C62.41 "IEEE Recommended Practice on Surge Voltages in Low-Voltage AC Power Circuits"

519 "IEEE Recommended Practices and Requirements for Harmonic Control in Electric Power Systems"

Appendix B
EGSA Standards
100 SERIES STANDARDS
Under the purview of the Power Generation Components Subcommittee.

EGSA 100B-1997—Performance Standard for Engine Cranking Batteries Used with Engine Generator Sets

Contains requirements for rating, classifying, applying, installing and maintaining engine cranking batteries.

EGSA 100C-1997—Performance Standard for Battery Chargers for Engine Starting Batteries and Control Batteries (Constant Potential Static Type)

Contains requirements for voltage and temperature limits, application and accessories for charging engine cranking batteries.

EGSA 100D-1992—Performance Standard for Generator Overcurrent Protection, 600 Volts and Below

Contains performance specifications for circuit breakers, field breakers, thermostats, thermistors and other temperature detectors.

EGSA 100E-1992—Performance Standard for Governors on Engine Generator Sets

Contains classifications, performance requirements and optional accessories for generator set engine governors.

EGSA 100F-1992—Performance Standard for Engine Protection Systems

Contains performance specifications for engine control systems including temperature, level, pressure and speed sensing.

EGSA 100G-1992—Performance Standard for Generator Set Instrumentation, Control and Auxiliary Equipment

Contains requirements for generator set engine starting controls, instrumentation and auxiliary equipment.

EGSA 100M-1992—Performance Standard for Multiple Engine Generator Set Control Systems

Contains performance requirements for manual, automatic fixed sequence and random access generator set paralleling systems.

EGSA 100P-1995—Performance Standard for Peak Shaving Controls

Contains requirements for parallel operation and load transfer peak load reduction controls.

EGSA 100R-1992—Performance Standard for Voltage Regulators Used on Electric Generators

Contains application and performance requirements for generator voltage regulators.

EGSA 100S-1996—Performance Standard for Transfer Switches for Use with Engine Generator Sets

Contains classifications, applications and performance requirements for transfer switches for emergency and standby transfer switches.

EGSA 100T-1995—Performance Standard for Diesel Fuel Systems for Engine Generator Sets with Above Ground Steel Tanks

Contains application and performance requirements for diesel fuel supply systems with above ground steel tanks for diesel engine driven generator sets.

101 SERIES STANDARDS
Under the purview of the Power Generation Systems Subcommittee.

EGSA 101G-1994—Glossary of Electrical and Mechanical Terminology and Definitions

Contains definitions of terms specific to the on-site power industry.

EGSA 101N-1992—Standard Nameplate Design for Engine Generator Sets

Guidelines for the information that should be included on the nameplate of a generator set.

EGSA 101P-1995—Performance Standard for Engine Driven Generator Sets

Contains classifications of use, prime mover configuration and ratings, and performance requirements for complete generator sets.

EGSA 101S-1995—Guideline Specification for Engine Driven Generator Sets, Emergency or Standby

Guideline specification in blank form for preparing specifications for emergency or standby generator sets.

EGSA 109C-1994—Code Listing: Safety Codes Required by States and Major Cities

A listing of national and international codes and standards adopted by U.S. states and selected major cities.

Referenced Code Organizations
The National Building Codes (NBC)
Building Officials and Code Administrators International (BOCA)
4051 West Flossmoor Road
Country Club Hills, IL 60478-4981
(708) 799-2300

The Uniform Building Codes (UBC)
International Conference of Building Officials ICBO)
5360 South Workman Mill Road
Whittier, CA 90601
(213) 699-0541

The Standard Building Codes (SBC)
Southern Building Code Congress International (SBCCI)
900 Montclair Road
Birmingham, AL 35213
(205) 592-1853

The Hartford Steam Boiler Inspection and Insurance Company
One State Street
P.O. Box 5024
Hartford, CT 06102-5024
Tel: (860) 722-1866
Fax: (860) 722-5106

ASME International
Three Park Avenue
New York, NY 10016-5990
800-THE-ASME (U.S./Canada)
E-mail: ASME InfoCentral

American Society of Heating, Refrigerating and Air-Conditioning Engineers, Inc. (ASHRAE)
1791 Tullie Circle, NE
Atlanta, GA 30329-2305

American National Standards Institute (ANSI)
11 West 42nd Street
New York, NY 10036

American Society for Testing and Materials (ASTM)
1916 Race Street
Philadelphia, PA 19103-1187

Federal Specifications (FS)
General Services Administration
7th & D Streets
Specification Section, Room 6039
Washington, DC 20407

National Fire Protection Association (NFPA)
Batterymarch Park
Quincey, MA 02269

Underwriters Laboratories Inc.
333 Pfingsten Road
Northbrook, Il 60062-2096

Electrical Generating Systems Association (EGSA)
1650 S. Dixie Highway, 5th Floor
Boca Raton, FL 33432

Institute of Electrical and Electronics Engineers (IEEE)
445 Hoes Lane
P.O. Box 1331
Piscataway, NJ 08855

Intertek Testing Services (ITS) Canada Ltd.
1055 West 14th Street, Unit 600
North Vancouver, BC, Canada V7P 3P2
Tel: (604) 984-4231
Fax: (604) 986-1205
Email info@its-pkb-vancouver.com

Appendix C
Microturbine Power Conversion Technology Review

R. H. Staunton
B. Ozpineci
Oak Ridge National Laboratory

1. INTRODUCTION

In this study, the Oak Ridge National Laboratory (ORNL) is performing a technology review to assess the market for commercially available power electronic converters that can be used to connect microturbines to either the electric grid or local loads. The intent of the review is to facilitate an assessment of the present status of marketed power conversion technology to determine how versatile the designs are for potentially providing different services to the grid based on changes in market direction, new industry standards, and the critical needs of the local service provider. The project includes data gathering efforts and documentation of the state-of-the-art design approaches that are being used by microturbine manufacturers in their power conversion electronics development and refinement. This project task entails a review of power converters used in microturbines sized between 20 kW and 1 MW.

The power converters permit microturbine generators, with their non-synchronous, high frequency output, to interface with the grid or local loads. The power converters produce 50- to 60-Hz power that can be used for local loads or, using interface electronics, synchronized for connection to the local feeder and/or microgrid. The power electronics enable operation in a stand-alone mode as a voltage source or in gridconnect mode as a current source. Some microturbines are designed to automatically switch between the two modes.

The information obtained in this data gathering effort will provide a basis for determining how close the microturbine industry is to providing services such as voltage regulation, combined control of both voltage and current, fast/seamless mode transfers, enhanced reliability, reduced cost converters, reactive power supply, power quality, and other ancillary services. Some power quality improvements will require the addition of storage devices; therefore, the task should also determine what must be done to enable the power conversion circuits to accept a varying dc voltage source. The study will also look at technical issues pertaining to the interconnection and coordinated/compatible operation of multiple microturbines.

It is important to know today if modifications to provide improved operation and additional services will entail complete redesign, selected component changes, software modifications, or the addition of power storage devices. This project is designed to provide a strong technical foundation for determining present technical needs and identifying recommendations for future work.

2.   POWER CONVERSION DESIGNS

This section considers the high-speed generator designs that are used in microturbine systems and the power electronics (i.e., power converter) that generally interface with the generators to develop the necessary 3-phase, line-frequency voltages.

2.1   Microturbine Generators
The highest efficiency operating speeds of microturbines tend to be quite high, often exceeding 100,000 rpm. The speeds are generally variable over a wide range (i.e., from 50,000 rpm to 120,000 rpm) to accommodate varying loads while maintaining both high efficiency and optimum long-term reliability. The microturbine drives a high-frequency generator that may be either synchronous or asynchronous (or non-synchronous). The caged rotor design in asynchronous (or induction) generators tends to make it a less-costly alternative to synchronous generators. Synchronous generators contain a magnetic rotor that is designed to use either rare earth permanent magnets or coils with additional hardware for delivering current (e.g., slip rings, brushes). Although asynchronous generators are somewhat rare in the industry, they are the generator of choice in wind and hydro generation applications.

Power requirements to the generator vary depending on the design. A synchronous generator with a wound rotor assembly will require dc power for energizing the rotor poles. An asynchronous generator in most microturbine applications will require a 3-phase current to the stator at a frequency correlated well to the rotational speed so that power is produced.

Ingersoll Rand's microturbine makes use of a high-speed gearbox instead of a high-frequency power converter.

In conventional applications, synchronous generators have an advantage where they can be connected directly to the grid if speed is properly regulated. This is generally* not the case in high-speed microturbine applications. For all generator types, a 3-phase, high frequency voltage, typically in the range of 1,000 Hz to 3,000 Hz, will be developed that must be converted to line frequency before the generated power becomes usable.

2.2   Power Converter Design

Figure C-1 shows a general diagram for a microturbine generator system followed by a power converter and a filter. The ac/ac power converter essentially converts high frequency ac to 50 or 60 Hz ac.

*An exception will be seen later where one manufacturer chose to use a conventional low speed generator after gearing down the turbine speed.

Figure C-1. General microturbine diagram.

The power converter can also be designed to provide valuable ancillary services to the power grid or microgrid. These services may include voltage support, sag support, static volt-amp-reactive (VAR) compensation, load following, operating reserve (e.g., spinning or non-spinning), backup supply, and/or start-up power for the microturbine or other local microturbines. Voltage support is common for griding-dependent operation while load following is used for grid-connected operation. Operating reserve capability may or may not be recognized by the local electricity provider depending on their current tariffs and the capabilities of the microturbine installation. The availability of backup supply and start-up power varies not only by microturbine manufacturer but also by what options may be purchased with the microturbine. For this reason, it will become a topic of discussion in contacts with manufacturers (see Sect. 3.3).

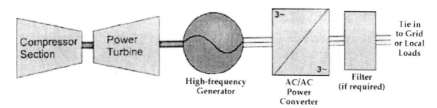

Figure C-1. General microturbine diagram.

2.2.1 DC Link Converter

The most common power converter topology that is used for connecting microturbines to the grid is the dc link converter. Figure C-2 shows a microturbine generator feeding power to an active rectifier circuit (or, alternatively, a passive rectifier) followed by a dc link and inverter circuit.

The high frequency power from the generator must be converted to dc before the inverter can reconstruct a three-phase voltage supply at lower frequency required for grid connection. A controller manages the operation of the active rectifier and inverter circuitry by ensuring that functions such as voltage following, current following, phase matching, harmonic suppression, etc. are performed reliably and at high efficiency. The controller may be mostly on-board, pc-based, a processor linked to a pc, etc., depending on constraints and factors such as desired microtur-

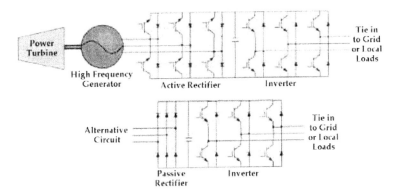

Figure C-2. Simplified diagram of a dc link converter.

bine packaging, desired versatility, type of available features, and the sophistication/maturity of the system design.

2.2.2 High frequency link converter

Another type of power conversion circuit that is of high interest is the high frequency link converter (HFLC). Figure C-3 shows a microturbine generator feeding 3-phase power to a rectifier and the dc is then fed to a high frequency, single-phase inverter so that a compact, high frequency transformer can be used. The secondary of the transformer feeds an ac/ac converter that takes the single phase, high frequency voltage to produce a 3-phase voltage at a frequency and phase needed to make a direct connection to the grid. Although the HFLC requires a higher part count, the circuit provides several advantages including:

- The use of a transformer for robust isolation

- The high frequency inverter permits the use of compact, high-frequency transformers

- The use of a transformer permits the easy addition of other isolated loads and supplies via additional windings and taps

- The circuit eliminates the need for static transfer switches

- Ancillary services can be provided with control software changes and additional hardware

- Adding additional hardware is easier

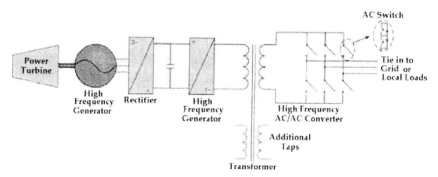

Figure C-3. Simplified diagram of a high frequency link converter.

Thus, a well-designed HFLC that is controlled by software could potentially provide unique characteristics (e.g., additional voltages, isolation/protection) to the microturbine owner. The system may offer certain advantages for growing with the needs of the owner. No microturbine manufacturer is presently marketing generation systems using an HFLC.

The data gathering effort will try to identify any development efforts or other experimental programs involving HFLC or any other unique or innovative power converter topologies.

2.2.3 Cycloconverter

A cycloconverter or a matrix converter could be used to connect the microturbine generator to the grid instead of using a rectifier and an inverter. These converters, as shown in Figure C-4, directly convert ac voltages at one frequency to ac voltages at another frequency with variable magnitude. For this reason, they are also called frequency changers. The disadvantages of these converters are that they have double the number of switches compared to the dc link approach and they do not have a dc or ac link to store energy. Without energy storage in the converter, any fluctuations at either side of the converter will directly influence the other side. In addition to this, it is not possible to connect a battery or any other power source to these converters unlike the dc link converter or the HFLC.

A cycloconverter can still be used for microturbines with the high frequency link inverter. Instead of converting the generator voltage to dc and then to high frequency ac, a cycloconverter can directly convert the three-phase ac voltage to single-phase high frequency ac voltage.

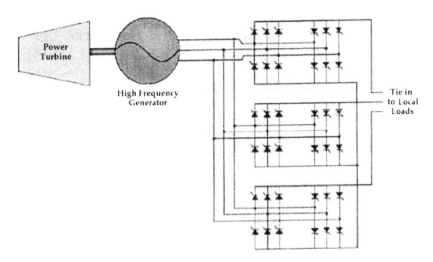

Figure C-4. Simplified diagram of a cycloconverter.

3.   INFORMATION OBTAINED FROM
      INDUSTRY AND TECHNOLOGY REVIEW

The data gathering effort performed in this study was conducted from December 2002 through March of 2003. Information was obtained through a variety of means including Internet searches, inspections of microturbines, review of microturbine manuals, and conversations with company engineers.

3.1   Information Needs

Questions were sent to microturbine power converter manufacturers and developers from several companies. Examples of the questions are provided in Appendix A along with descriptive text explaining precisely what information was being sought. This section lists the types of information the data gathering effort was designed to obtain.

The types of information that were sought during the data gathering effort include the following:

Generator Type and General Description
•   Asynchronous vs. synchronous
•   Packaging – power converter location

- Turbine speed range
- Modes of operation (e.g., stand alone, grid connect)
- Power rating
- Cost of power converters
- Manufacturer/supplier identification

Power Conversion (technical)
- Determine if the converters are pulse-width-modulated or if they use a commutated-pulse architecture (i.e., line commutated inverter)
- Identify type of circuit topology
- Determine the switching frequency
- Determine internal circuit control (onboard microprocessor or a computer with software)
- Determine if there are concerns with electromagnetic interference (EMI) and harmonic distortion
- Types of ancillary services/special features provided by power converter
- Required accessories
- Other features

Component issues
- Determine if the architecture uses MOSFETs vs. IGBTs or pn diodes vs. Schottky Diodes
- Determine how close the switching devices operate to their maximum ratings
- Identify the operating/maximum temperatures
- Identify the heat removal method
- Determine how much fault current may be developed and for how long

Analysis – Determine whether hardware changes would be needed to expand the ability of the system to provide grid support (i.e., ancillary services) or if they can be accomplished with just a change to the processor or software.

As with many inquiries sent to industry, a rapid and enthusiastic response is a rare exception. Engineers, marketing, and sales personnel are generally overworked and unable to devote time to preparing 7 responses, even if they see some potential benefit in doing so. However, this data gathering effort had an additional challenge in that the industry it sought to query is young, struggling for survival, addressing numerous urgent problems, and very sensitive to releasing technical design-related information. In addition, the early production microturbines now in service, including mechanical systems and power converters, have experienced reliability problems. Although the contacts made with manufacturers were not designed to probe for such information, responders may somehow feel that full admission of such matters was expected. One power converter manufacturer, after considering the information request, simply declined to respond without offering any explanation. In spite of these challenges, reasonably complete results were eventually obtained by perseverance, seeking out alternate contacts from companies, inspecting local production microturbines, and reviewing literature (i.e., including owner's manuals and training manuals).

3.2 Manufacturers of Power Converters

This section presents basic information relating to domestic and foreign power converter manufacturers whose products are, or can be, used in microturbines in the size category from 20 kW and 1 MW. This study will, in later sections, focus primarily on domestic manufacturers; this is necessitated by the fact that detailed design data, which is quite difficult to obtain from U.S. manufacturers, will be more difficult (i.e., essentially impossible) to obtain from foreign manufacturers.

It becomes very evident just how *new* the microturbine industry is when one considers all of the manufacturers that are either marketing their products as of the last few years, are taking orders for soon-to-be introduced products, or are strictly in the development of products. Even the "oldest" manufacturers are just beginning to see their products arrive at the first major overhaul point. At best, the reliability calculations for microturbines and the associated power converters are now becoming less theoretical and more based on actual in-service data.

Table C-1 provides a list of all known power converter manufacturers and/or microturbine manufacturers. It indicates their relevant product or development activity, whether their power converters are an in-house design and production effort, and additional notes or comments. The table indicates that a number of microturbine manufacturers do not produce their own power converters for a number of reasons. For instance, Elliot Energy Systems purchases all of the power converters used in their microturbines from Bowman Power Systems, and Ingersoll Rand Energy Systems uses a gearbox that enables them to use conven-

Figure C-5. Capstone microturbine - 60 kW.

tional induction and synchronous generators that connect to the grid/loads without the use of a power converter (see Sect. 3.3.2). Although the parent company of Elliott Energy Systems, Ebara Corporation of Japan, is reportedly becoming a new supplier of power converters to Elliott, it is too early to report on how these power converters might differ from those supplied by Bowman.

Turbec produces a Model T100 105 kW combined heat and power (CHP) microturbine that uses a dc link power converter to convert the generator's high frequency output to useful power. Turbec markets primarily to Europe with distributors in Italy, Switzerland, France, UK, Ireland, Denmark, and Germany; therefore, this study's consideration of Model T100 is somewhat limited. However, several unique features/characteristics that merit consideration are briefly summarized in Sect.

Table C-1. Power converter and/or microturbine manufacturers whose products are in production and available on the market

Manufacturer	Product or Development Activity	Producer of power converters[a]	Notes
Ballard	Ecostar™ power converters from 10 kW to 1 MW	Yes, for other generation applications	Fuel cells are Ballard's main product line
Bowman Power Systems	Turbogen™ family of microturbines ranging from 25 kWe to 80 kWe	Yes	Supplies converters to Elliot Energy Systems, Inc.
Capstone Turbine Corporation	30 kW and 60 kW microturbines (200 kW microturbine under development)	Yes	Integral module; DOE-AMP b participant
Cummins	30 kW and 60 kW microturbines	No	Basic microturbine systems are obtained from Capstone
Elliott Energy Systems, Inc. owned by the Ebara Corporation (Japan)	35 kW, 60 kW, and 80 kW microturbines, products have used Bowman power converters; however, the Ebara Corp. is an emerging supplier	No	Supplies mechanical microturbine systems to Bowman Power Systems
Ingersoll Rand Energy Systems	70 kW PowerWorks microturbine (no power converter used), larger units planned.	NA (see text)	DOE-AND participant, gearbox is used to reduce rpm and to facilitate use of low speed generator.
Turbec AB (owned by ABB & Volvo)	Model T 100 is a 100 kW microturbine marketed almost exclusively to Europe	Yes	Modest sales in U.S.; not fully included in the present study
Xantrex	A wide range of power converters are available for any type of generator. applications	For other generation	No present microturbine applications

(a) Indicates whether the power converters are an in-house product.
(b) DOE-AMP = Department of Energy's Advanced Microturbine Program, not all participants are involved in development of power electronics and therefore not all are listed.

Figure C-6. Capstone (30 kW) opened to show major assemblies.

3.3.

Table C-1 also lists some manufacturers, such as Ballard and Xantrex, who market versatile power converters designed for use with renewable energy generators (e.g., wind, solar) and other types of distributed generation applications. These applications are distinct primarily because they need not accept a high-frequency output from generators such as used in microturbines. Although Ballard and Xantrex have also developed products capable of converting power from high-speed microturbine generators, they are not currently marketed.

Table C-2 lists the companies that are currently developing a microturbine and/or power converter(s). In some cases, the companies are pursuing almost 100% product development research activity with technology demonstrations planned within a year. Although the companies listed in this table do not presently market specifically microturbine power converters, most appear to be quite notable for the high degree of innovation they are using to attain versatile, new products.

Table C-3 provides the Internet addresses for the full list of vendors presented in this section.

3.3 Data Provided by Manufacturers

This section provides information on power converters provided by five primary manufacturers of power converters. The information will include a summary of general features, operating parameters, ancillary services, special features, power quality, circuit topology, and other circuit details including component types and thermal specifications. A short summary of industry needs is also provided. The second subsection will consider important companies also meriting special consideration because of the unique features of their design and/or because they offer an alternative design approach.

3.3.1 Primary manufacturers

Based on the information presented in Section 3.2, the final list of companies in the U.S. market who are presently producing and marketing power converters suitable for distributed generation (i.e., not *just* microturbines) is quite small. It includes the following:

- Ballard (provided power converters for Honeywell microturbines in 1998)
- Bowman Power Systems
- Capstone Turbine Corporation
- Xantrex (no presently marketed microturbine application)

General Electric (GE), which is well along in their development of a relatively large microturbine, was also helpful in providing information and is included in this data summary. Ballard has developed a 110 kVA "microturbine system" (i.e., power converter system) that the company says is capable of operating in grid and stand-alone mode with all the grid connect and retry strategies in accordance with IEEE 1547.* Xantrex, which has developed a wide range of power converters that can be used

---

*IEEE 1547 is the *Standard for Interconnecting Distributed Resources with Electric Power Systems* (See Sect. 4.1.2.)

Table C-2. Potential power converter and/or microturbine manufacturers who are now in the R&D phase of product development

Manufacturer	Product or Development Activity	Producer of power converters[a]	Notes
AeroVironment	No firm microturbine project yet, has Power Electronics Module (PEM)	Yes	Solar products, iPower Technologies
DTE Energy Technologies	ENT 400 kW microturbine to be added to internal combustion generator line	No	Power converter will be purchased from Turbo Genset
General Electric, Global Research Center & GE Industrial	Developing microturbine for 175 kW (ultimately, 175-350 kW) and power electronics	Yes	DOE-AMP participant
Northern Power Systems	Research into development of advanced algorithms and modular circuit designs; may enable seamless transitions	Plans to produce power converters	Nothing announced yet to industry prior to completing work on patents.
SatCon Power Systems	Developing a wide range of power conversion systems for all types of distributed generation	Unknown	Storage systems also in development.
Turbo Genset	Developing a 50 kW microturbine and power converter system	Yes	Microturbine will be sold to DTE Energy

(a) Indicates whether the power converters are a product of their in-house design efforts.
(b) DOE-AMP = Department of Energy's Advanced Microturbine Program, not all participants are involved in development of power electronics and therefore not all are listed.

Table C-3. Manufacturers and internet URLs

Manufacturer	*Internet URL*
Ballard	http://www.ballard.com/
Bowman Power	http://www.bowmanpower.com/
Capstone Turbine Corporation	http://www.microturbine.com/
Cummins Northwest Inc.	http://www.cumminsnorthwest.com/PowerGen/Microturbine.asp
DTE Energy Technologies	http://www.dtetech.com/
Elliott Energy Systems, Inc./Ebara Corp.	http://www.elliott-turbo.com/new/products-microturbines.html
General Electric (GE)	http://www.eren.doe.gov/der/microturbines/pdfs/geslide.pdf
Ingersoll Rand Energy Systems	http://205.147.212.185/
Northern Power Systems	http://www.northernpower.com/
SatCon Power Systems	http://www.inverpower.com/products/alten/alten.html
Turbec AB (owned by ABB & Volvo)	http://www.turbec.com/
Turbo Genset	http://www.turbogenset.com/
Xantrex	http://www.xantrex.com/Products/index.asp

in microturbine applications, has also provided information.

The investigation of Capstone power converters was performed using (1) responses from the company, (2) information obtained through visits to three research sites** where Capstones are used, and (3) a review of Capstone documentation. All of the above companies were asked in a series of questions to provide details pertaining to (1) their microturbines (or the microturbines from other manufacturers that make use of their power converter), (2) technical aspects of the power converters, (3) technical issues (e.g., power quality), and (4) electronic component details.

Table C-4 provides information on the microturbine generator type, packaging, power converter switching frequency, features and manufacturer. The most detailed information is listed under the two current manufacturers of microturbine power converters, Bowman and Capstone. Detailed information for GE's product was also obtained because they are well along in developing a specific microturbine product. Clearly among the manufacturers, power converter designs are being directed for use with synchronous generators although Xantrex indicates that they are also willing to adapt designs to non-synchronous machines. Current power converter products are modular and integrated into the microturbine; however, for GE's 175 kW design, external packaging of the converter is planned. Power converter switching frequency is generally up to 8 kHz for the different manufacturers, although Xantrex designs may use much higher frequencies.**

A key feature indicated in Table C-4 is that all of the manufacturers use some type of reprogrammable digital control system for the power converters. This provides a potentially needed level of versatility that may become critical in the future (see Sect. 4.1). The modes of operation generally include stand alone (i.e., for supplying power to local loads only) and grid connect, which must be synchronized to the grid. The GE microturbine is planned for grid-connect operation only. Switching between modes is a major issue. The quality or speed of the transitions from one mode to another varies from manufacturer to manufacturer and also depends on the mode to which the unit is being switched (see Sect. 4.1). The transitions generally cause a voltage interruption. This can

---

**Project staff inspected a 30 kW Capstone microturbine at the CHP Integration Laboratory – A National User Facility at ORNL, a 60 kW Capstone at the High Temperature Material Laboratory (HTML) at ORNL, and a 30 kW unit at the EPRI PEAC Corporation in Knoxville, Tennessee.

**Higher switching frequencies generally come with a power de-rating penalty.

Table C-4. Microturbine manufacturer, operating parameters, and general features

Subject	Ballard	Bowman and Elliott*	Capstone Turbine Corporation	General Electric	Xantrex
Type of generator used in microturbine	Synchronous	Synchronous	Synchronous permanent magnet generator	Synchronous permanent magnet	May be used with synchronous or non-synchronous
Packaging of overall unit	Modular circuit	Integral packaging	Integral packaging	Power converter in nearby cabinet	Not currently integrated in package
Turbine speed range	Customer dependent	68,000 rpm (normal)	45K-96K rpm	12K -50K rpm	NA
Power converter switching frequency	4-8 kHz	8 kHz	Unknown	=8 kHz rectifier, =5 kHz inverter	2-18 Ez depending on power level
Modes of Operation and seamlessness of transfer	Stand-alone & grid-connect (seamless switching possible)	Stand-alone & -grid-connect with sub-second switchover	Stand-alone & grid-connect with auto-switching (see text)	Grid connect only	Stand-alone & grid-connect (interruption only when grid V drops)
Type of digital control	Microprocessor (reprogrammable)	Microprocessor (reprogrammable)	Digital signal processor	Flash-based microprocessor	Digital signal processor by Texas Instruments
Accessories for operation in different modes	Unknown	Optional dual-mode switch required for grid independent mode	Battery required for stand-alone mode and black start (see text for other accessories)	NA	Uninterruptible power supply required for black start of microturbine
Power rating of power converter	10 kW to 110 kW	60 kW and 80 kW	30 kW and 60 kW (200 kW future)	175 kW @ 40°C	Power converters from 5 kW to 1 MW
Manufacturer of mechanical portion	No microturbine product or application at present	Elliot Energy Systems	Capstone	General Electric	No microturbine product or application at present
Manufacturer of power converter	Ballard Power Systems	Bowman Power Systems	Capstone	General Electric	Xantrex
Cost of power converter and/or microturbine	Depends on application	—	~$30,000 for the 30kW model microturbine	TBD when product is introduced into market	$70/kW to $ 1,000/kW for power converters depending on application

Power converters supplied by Bowman Power Systems

be a significant problem in many of today's applications in industry and in the commercial sector. Only Ballard claims to have completely seamless transitions; however, this must be demonstrated in a microturbine system and made available in the commercial market.

The Capstone microturbine has stand-alone and grid connect modes with automatic switching between them. Transferring from grid connect to stand alone takes 2 to 4 minutes while returning to grid connect takes only 5 seconds.

The Capstone can be used with other accessories than the battery listed in the table. For instance, Capstone's Dual Mode Controller contains a utility disconnect and will allow automatic switching between grid connect and stand alone modes. Up to 20 units can be MultiPacked (paralleled) as a standard feature, and up to 100 can be MutiPacked using the optional Capstone Power Server. The purchase of an external power meter will facilitate operation with reverse power protection and load following.

Table C-5 lists the various types of ancillary services that the power converters are capable of supplying to the grid. The listing also includes a few special features of interest such as operation by remote control and monitoring device and communications that support operation of multiple microturbines in a parallel configuration. A review of Table C-5 shows that, with the exception of Bowman/Elliott and GE, the power converter manufacturers claimed an extensive list of ancillary services and special features. The analysts strongly suspect that in some cases, claims may have been made on a more theoretical basis (i.e., making claims of *potential* capabilities). Therefore, some level of caution is advised in considering the implicit claims indicated by the table. However, this is essentially a moot point for the purposes of this study, since the primary goal of the study is to determine exactly this—the *potential* features of the microturbine power converters.

As indicated above, the Capstone load following service and reverse power protection feature require the use of an accessory (i.e., an external power meter called the "pulse issuing power meter").

Table C-6 shows a summary of the microturbine power converter topology, the type of switching components used, key thermal specifications, cooling method, and power quality concerns (if any). For all manufacturers, the table indicates many similarities including a dc link topology, pulse width modulation (PWM) waveforms in the inverter output, and some type of forced air heat sink arrangement (except for

Table C-5. Microturbine ancillary services and special features

Subject	Ballard	Bowman and Elliott*	Capstone Turbine Corporation	General Electric	Xantrex
Voltage sag support	Yes		Possible		Yes
Static VAR compensation	Yes	Can select 1.0 to 0.6 leading/lagging power factors (Elliot)	Possible	Yes—power factor control only	Yes
Load following		Yes	Yes		Yes
Operating reserve (spinning or non-spinning)			Yes		Yes
Backup supply	Yes		Yes		Yes
Blackstart capability (without external grid)	Yes	Greystart available provided gas is present and oil is warm	Yes	Yes	
Remote controllability	Yes	Yes	Yes	Yes	
Communication (operation in parallel with other units)	Yes - has multi-unit capability for up to 10 units in parallel.	Yes	Yes, using the standard MultiPac RS485 communications port	Yes	
Others		Reverse power protection (in Elliot microturbine therefore an assumed feature of Bowman)	Peak shaving, Reverse power protection, and auto-restart (following a fault)		

*Power converters supplied by Bowman Power Systems

Ballard, which uses liquid cooling). Certain topics, such as how close the switching devices operate to their maximum ratings, proved to be too sensitive to result in any significant response from the companies. The filtering referred to in the bottom two rows of the table has proved effective for all manufacturers in producing an output waveform that does not produce EMI, sinusoidal in shape, and often with lower harmonic distortion than found on the local grid.

Regarding the type of circuit topology, GE provided additional details regarding their 175 kW power converter. The 3-phase, two-level active IGBT power electronic bridge converts the 3-phase, high-frequency generator voltage to dc. The voltage on the dc link is regulated by the active rectifier and it feeds a 480V, 3-phase, IGBT inverter. The dc link voltage is also monitored so that turbine loading/speed can be adjusted according to the apparent inverter loading of the dc link. 15

GE also supplied additional details regarding how close the switching devices operate to their maximum ratings. Obviously, this depends on the output voltage of the generator and other factors such as load and speed. GE follows general design rules of 10% voltage margin for the main power electronics. This may seem to be a tight margin but, in reality, how appropriate it is can be determined only from the design details.* GE is still developing the overall circuit control strategy and will be operating the power conversion electronics for the first time in the spring of 2003. GE will be assessing device stress as well as other system operating points, and their test results will determine whether the bridge is correctly used and rated. GE states that sophisticated junction temperature techniques are used to maintain device temperatures within published limits.

The final question asked of manufacturers was what they would

---

*GE is currently using 1400V parts with about a 900V dc Link. GE attempts to apply 10% extra margin after all worst case control and transient effects have been accounted for. Hence, device switching transients will be added to the 900V link voltage before the margin is applied. The switching transients can be controlled by altering the gate drive characteristics and stray bus inductance.

It is possible to require less margin if higher performance bridge topologies (buswork) and controls are used. Manufacturers apply margin based on their experiences, control complexity and response (i.e., ability to control the dc Link Voltage during input/output transients) and power electronic bridge design. It is very advantageous for an equipment manufacturer to achieve a higher rating or extra performance over a competitor's product by developing their own philosophy for rating and application capability.

Table C-6. Microturbine circuit topology, components, thermal specifications, and power quality concerns

Subject	Ballard	Bowman and Elliott*	Capstone Turbine Corporation	General Electric	Xantrex
Identify circuit topology	DC link converter	DC link converter	DC link converter (current regulated)	DC link converter (see details in text)	Voltage source dc: link converter
Identify type of waveform generated	Pulse width modulation (PWM)	PWM	PWM	PWM	High frequency sine - triangle PWM
Identify type of power devices in inverter	IGBTs	IGBTs or intelligent power modules (IPM) by Semikron/IR	IPM using IGBTs	IGBTs manufactured by Powerex	IGBTs or IPM by Eupec, Powerex
Switching devices operate how close to maximum ratings?	No information provided	Confidential	No information provided	Dependent on factors such as load and speed (see text)	No information provided
Maximum ambient T	40°C	45°C	50°C	40°C without derating	45°C (min. is -20°C)
Maximum device junction temperature	Unknown	150°C	115°C	125°C	150°C (not operated above 110°C)
Method of cooling of power switching components	Water/liquid cooling	Force air cooling with heatsink	Force air cooling with heatsink	Force air cooling with heatpipe heatsink	Force air cooling with heatsink *or* water cooled
Permissible overload current	No information provided	300% of peak rated for 2 cycles	200% for 1s, 150% for 10s, 125% for 30s, and 110% for 60s	This design parameter is TBD	Depends on application
Filtering and shielding	Both are used	Both are used	Both are used	Filtering included for harmonics and EMI	Both may be used
Power quality considerations	No EMI or harmonic distortion problems	No EMI or harmonic distortion problems	No EMI or harmonic distortion problems. Harmonics well below IEEE 519 limits.	Harmonic levels will meet IEEE 519	No EMI or harmonic distortion problems

*Power converters supplied by Bowman Power Systems

like to see from the Government or industry (e.g., product, component) that would be of help to them. The responses shown in Table C-7 were obtained from Ballard, GE, Capstone, and Bowman Power Systems.

Table C-7. Present industry needs from the perspective of power converter manufacturers

Manufacturer	Needs from the government and/or industry
Ballard	Ballard states that they would like to see funding for high-voltage/current silicon carbide diode development.
Bowman and Elliott	Bowman Power Systems indicated that it is desirable to have a sustained incentive, either by way of a fund or a geographic-supported region, which developers could rely upon over a number of years to provide a proving ground for the technology.
Capstone Turbine Corporation	Capstone indicated that they would like to see continued development and adoption of interconnection standards.
General Electric	GE stated that they would like to see the implementation of codes and standards to improve and accelerate technology acceptance.
Xantrex	No response

3.3.2 Alternative design approaches of interest

Turbec AB's mostly European product has a number of interesting features worth reviewing. Their T100 microturbine is rated at 105 (±3) kW, has a nominal speed of 70,000 rpm, and is designed for operation indoors only. The unit provides CHP with a gas-water, counter-current flow heat exchanger providing a 122°F to 158°F temperature rise. The net electrical efficiency is 30% while the net total efficiency is 78% (assuming full use of the heated water). A two-pole permanent magnet high-speed generator feeds 500 VAC, 2333 Hz to a dc link converter that produces

400 VAC, 3-phase, 50 Hz or alternatively 480 VAC, 60 Hz. The converter can be operated in reverse to use the generator as a starter motor for the microturbine. A power module controller (PMC) controls and operates the entire electrical system. The T100 features a communications connection that permits the operator to monitor and control the microturbine from a remote location.

It would be unfair to consider the "challenge" of power converters seamlessly transitioning before and after grid outages without also considering how one alternative microturbine approach already performs this function with relative ease. Ingersoll Rand's synchronous generator, which uses a high-speed gear box instead of a power converter, features a switchgear package that includes two motorized breakers that act to isolate segmented loads from either the microturbine or the facility distribution system. One breaker is used to connect the microturbine when it is synchronized. The other, upstream breaker acts to isolate the microturbine/segmented load combination in an intentional island. When the electrical protection system detects a grid failure, the upstream breaker is automatically opened and the microturbine assumes all power man-

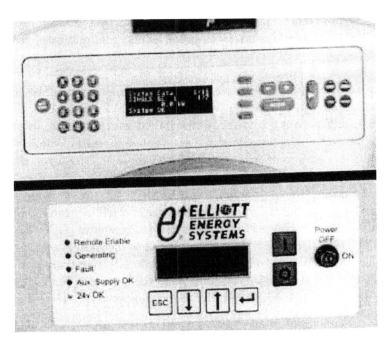

Figure C-7. Front control panels of Capstone and Elliott microturbines

agement to the loads. This transition occurs without any loss of power to the loads. When the protection system detects that the grid is back to normal, another seamless transition is made to return to grid parallel operation.

3.4   Notes on Other Connection Technology

Microturbines produced by companies such as Bowman, Elliott, and Capstone, can be operated in a standalone mode to supply local loads or connected directly to the grid. The internal circuitry provides effective filtering for EMI and harmonic distortion and ensures that the microturbine's 3-phase power is fed into the grid synchronously. There are various options for interfacing small power sources to the utility. This section describes one such solution, called the GridGateway.

This 240V (up to 7kW) residential interconnection system, which consists of two DSP-controlled double pole relays, acts as a transfer switch by connecting and disconnecting a distributed generation system to the grid and/or load. It senses the three-phase grid and generator voltages and disconnects the distributed generator in cases of under/ over-frequency, under/over-voltage, and over-voltage THD. The fault limits 19 for voltage and frequency can be set by the user through a graphical user interface on a computer, which communicates with the unit through a serial port.

The power to the control circuitry of the GridGateway system comes from both the generator and the grid; therefore, if either of them fails, the unit will continue to function. The system can operate in four states, utility only, no connection, distributed generator only, or parallel operation. When it is turned on, the distributed generator switch is open, isolating the generator from the load and the grid. However, the grid switch starts in the state it was last in, before power was turned off.

According to the manufacturer, EPRI-PEAC, this interconnection system can be connected to the power meter and will be ready to operate in half an hour because of its plug-and-play capability.

The GridGateway technology uses relays instead of solid state devices. A concern with relays is that they could potentially get stuck, which may cause undesirable problems to the generator, the load, or the grid. Tests need to be done in order to evaluate system performance and whether the reliability of the relays is adequate without additional protection in case of failures. If this system passes the tests fundamentally intact, it could be used as a transfer switch complementing the utility interface converter.

Figure C-8a. Elliott microturbine (80 kW)
opened for servicing

Figure C-8b. View inside front cover of Elliott

4.0   CONVERTER TECHNOLOGY AND RELIABILITY

This section summarizes findings from Section 3, discusses the present status of microturbine technology, describes various technical needs, and discusses aspects of system reliability from the perspective of a fledgling industry.

4.1   Status of Power Converter Technology

This section discusses microturbine operating modes and the voltage interruptions that accompany switching between modes, the present status of software used in the digital controllers, and the need for universal interfaces and standard communications. The final discussion focuses on the full range of ancillary services that microturbines can be made to provide. The section states how manufacturers should be *encouraged* to make various services available and how certain ones might be standardized.

4.1.1 Operating modes and transitions

The present, conventional microturbine designs, when in the stand-alone mode, act as a voltage source to local loads. That is, the digitally controlled power converters regulate voltage and frequency while supplying current as needed by the load. However, the designs do not regulate voltage well enough to share reactive power flow. The microturbines are all operated at essentially the same voltage. When the microturbines are in the grid-connect mode, they act as a current sources, that is, they follow the voltage and frequency of the grid, and regulate the current, or power output, to a preset value. That value can either be constant, or it can load follow if desired.

None of the microturbines that is presently available in the market provides seamless switching between operating modes. This may be acceptable early-on in this emerging industry, but it will not be acceptable in the long term if greater market penetration is to occur. At a minimum, adaptations should be made to the system designs so brief voltage drops during switching can be bridged reliably.

4.1.2 Software used in the programmable digital controllers

The digital control schemes for the controllers were found to be versatile in all cases. This was an encouraging finding since, given the necessary sensors and controls, the controllers can be reprogrammed to provide a number of new features and services to both the user and the grid. Given this situation, the study considered the present status of the software now in place.

Capstone, as an example, provides two systems that aggregates their microturbines and provides services. With both systems, the microturbine is either operated in the stand alone mode as a voltage source or in the grid connect mode as a current source. The power electronics hardware is capable of regulating both current and voltage simultaneously, but the control system is not presently programmed to perform this task. This is because the Capstone philosophy is to supply a product that meets IEEE 1547. IEEE 1547 states that the microturbine, or other distributed generation device, is not allowed to impact voltage. In reality, this may only limit a *group* of several microturbines operating in parallel assuming a reasonably "stiff" voltage in the feeder. Therefore, this standard requires that, in the grid connect mode, microturbines do not regulate voltage, they only follow the grid voltage. Thus, the microturbines cannot be used for distribution system voltage regulation at this time.*

Bowman's power converters operate not only with permanent-magnet, high-speed alternators (microturbines), but also with wind generators and fuel cells. The Bowman units presently provide voltage regulation and power factor control. However, when the units are in the grid connect mode, the inverters acts as a current source only, similar to the Capstone Unit.

4.1.3 Universal interface/communications

In researching the ability of microturbines to operate in parallel, it was found that there is generally some level of capability in this area without purchasing special interface or communications accessories. It was also apparent that, as the number of units that a customer wishes to operate in parallel increases, one or perhaps two different controllers are necessary. The question of compatibility of microturbines from different manufacturers working together in large numbers is, at best, uncertain due to the lack of any type of universal interface. An example of two necessary interconnection accessories from one manufacturer is provided

---

*For the same reason, the several microturbines operating together cannot be used for voltage support at the end of long feeders. The microturbines are also capable of supporting voltage by reducing load, but this also is would not be allowed.

below.

The first Capstone control system is the Capstone Remote Monitoring System (CRMS). The CRMS allows you to communicate locally with up to 40 microturbines. The microturbines are "daisy chained" together using their serial ports and an RS 232 ethernet connection. One turbine is the master and the rest are in the "slave" mode. Data provided from each microturbine includes control panels, strip charts, trend graphs, event alarms and automation panels. The load cycle and scheduler automatically starts, stops, and commands power to the micro turbines locally or remotely. The event monitor logs starts, stops, and faults. The turbines can be controlled to provide a set power level or to load follow. If they are load following, a compatible power meter must be purchased which provides a pulse count. The CRMS will load the turbines for the maximum fuel efficiency point. If a site has several Capstone turbines, the only additional need is for the software to run them with the CRMS.

The second system offered by Capstone is the Power Server CPS 100. The CPS 100 integrates up to 100 turbines into a single generation system with one point of control. The CPS100 contains a microprocessor which provides high speed networking via one RS-232 connection or a modem. There is also internet command via TCP/IP. There is an interface to the power meter so that control can be load following/peak shaving and dual mode (grid connect/stand alone). The CPS100 will load the turbines for the maximum fuel efficiency point. The CPS100 will also balance runtime so that all turbines are run at the same number of hours over time. There are also analog and discrete inputs for communications from other plant equipment.

4.1.4 Proposed requirements for ancillary services

As can be observed from the vendor data summarized in this report, there is not a standard on what electrical services the manufacturers provide. The analysts of this study believe that some of these services should be provided without extra accessories and others made available as options. Initially, manufacturers should be *encouraged* to make various services available. This may quickly lead to those same services being expected by potential customers.

Recommended or required services:
1.   Stand-alone and grid-connect operation modes and seamless switching between them in subcycles without voltage interruption.

2. Blackstart capability to start the microturbine without the external grid and whenever needed.
3. Remote controllability, so that the microturbine system can be controlled and monitored by a computer or a control panel without the need of actually being physically next to the system. This also brings the possibility of internet monitoring and control. It can also give access to the utility to control the system to prevent anti-islanding and to de-energize it remotely for maintenance and repair.
4. Onboard, reprogrammable microprocessors, so that the firmware can be upgraded easily, maybe even by the user, remotely.
5. Communication or the ability to operate with others so that numerous microturbines from the same or different manufacturers can be connected to each other and controlled by the master unit (one of the microturbines). As an extension to this point, with the cooperation of the converter manufacturers for other alternate energy sources, microturbine systems should be able to communicate with the other alternate energy systems.
6. Modularity or flexibility to add any additional hardware for optional services without taking everything apart.

Optional Services:
1. Voltage sag support
2. Static VAR compensation
3. Load following
4. Backup supply
5. Operating reserve (spinning or nonspinning)

The reason for including voltage sag support and static VAR compensation at an option level is that not all applications need these services; however, it is strongly recommended that these services be included since they can be implemented with just some additional software code and, if needed, they could contribute even with the microturbine off. Especially, considering the predicted possible increased importance of non-active power in the near future, static VAR compensation would be crucial option to have.

4.2 System Reliability in an Emerging Industry
This section provides an early assessment of reliability in domestic

microturbines. This assessment cannot yet be based on any type of quantitative evaluation, but rather, based loosely on certain industry events and product operating conditions that seem to be creating clear signals that are relevant to reliability.

4.2.1 Reliability issues

The microturbine industry is essentially new, where manufacturers are either marketing their products as of the last few years, are taking orders for soon-to-be introduced products, or are strictly in the development stage of new products. Even the "oldest" manufacturers are just beginning to see their products arrive at the first major overhaul point.

At best, the reliability calculations for microturbines and the associated power converters are now becoming less theoretical and more based on actual in-service data. However, a more realistic perspective shows subassembly designs that have been corrected for design weaknesses, often more than once, and field upgrades that alter any attempts at accurately tracking performance of the affected subassemblies.

Although this study was not intended to explore failures, multiple reports were heard concerning failures of power converters, EMI filters, and other systems. There are also many reports of aggressive actions taken by the manufacturers to correct problems in the field to maintain the highest degree of customer satisfaction and keep company reputation untarnished. The failures, the redesign process, and aggressive implementation of upgrades are all typical processes that take place when technologically complex products are introduced into the market.

Section 4.2.2 will explore some possible root causes for some of the failures that have been experienced in the microturbine industry to date.

4.2.2 Operating environment

In the course of this study, it was observed that the present microturbine manufacturers and vendors market their product to be located in outdoor locations where the microturbine cabinet and internal packaging (if any) are the only type of protection that the electronics have from the outside weather conditions.

The power converter electronics in the Capstone microturbines are packaged in a ventilated metal box located under the mechanical portion of the microturbine. The metal box is enclosed in the main cabinet of the microturbine. Air enters through a filter at the lower portion of the cabinet front through horizontal slits and travels a short distance before entering into the power converter box. The air is forced through the box with fans and exits at the opposite end. Some air flow can occur through these openings even when the microturbine is shut down. Such air flow would be partially impeded only by one or more air filters depending on the microturbine design. The Elliott microturbine design does not include the box located internal to the cabinet thus providing even less protection to the converter circuit.

The power converter electronics and digital control circuitry used for microturbines are both quite elaborate and sophisticated. It is unusual to see this type of electronics located in a semi-open cabinet located outdoors in the open because of how this might impact reliability. It is always a significant design challenge to have complex circuitry located in an uncontrolled, outdoor environment operate reliably. Looking throughout industry, examples of electronics routinely used in outdoor ambient conditions protected only by a cabinet/enclosure are few:

- Electronics used in vehicles
- Traffic light controls
- A portion of the electronics operated by electric utilities
- Portable electronics such as for communications

Although the above electronics are specifically designed to operate in severe ambient conditions, they are not fully valid comparisons. Many of these are better protected than in the microturbine example given above. For instance, a computer in an automobile may be enclosed in a nearly air-tight enclosure that would reduce the potential for condensation. Similarly, traffic control electronics may be enclosed in a box designed with a rubber seal. In these cases, the potential for condensation due to daily environmental cycles is significantly reduced.

In situations where microturbines are frequently shut down, left idle for hours or longer, and started up, there could be a potential for condensation inside the cabinet depending on changes in the ambient temperatures and humidity levels (e.g., daytime conditions vs. night conditions). Long idle periods during humid weather could cause moisture to be absorbed in certain components degrading insulating materials. In addition, some components (e.g., screw-down wire terminals) or cabinet surfaces may experience corrosion or become tarnished due to moisture.

The microturbine industry is not oblivious to these concerns. As indicated in Section 3.3, Turbec AB's T100 microturbine is designed for indoor installations only. Although it takes its air from an outdoor intake, that occurs only when the unit is operating (i.e., the power electronics would be warm precluding condensation). Additional design work and careful testing over time would be required to permit outdoor installations with greater exposure to the elements.

The fact that the microturbines are generally operated continuously outdoors is *helpful* from a reliability standpoint since component heating tends to keep surfaces dry and the ventilation prevents excess heat. The main concern during operating periods may be unusually hot, humid summer days; however, the use of conservative design margins and built-in thermal protection should prevent failures from occurring.

5.   SUMMARY AND RECOMMENDATIONS FOR FUTURE WORK

In the future, several ancillary services will be provided by a local, automated response, as opposed to system operator dispatch. The primary services that will be provided on a local basis include: reactive power supply and voltage control, black start, regulation, energy imbalance, network stability, spinning reserve, and load following. Each of these services requires that power be exported to the grid.

Gaining the ability to export power to the grid will require that the power electronics converter and interface control both voltage magnitude and phase angle. It would be necessary to control the electronics quite rapidly, in response to sensed local conditions. In addition, the capability to synchronize quickly will be needed. Finally, the power electronics would have to be capable of accepting power from a storage device such as a battery or ultra-capacitor because a microturbine is not capable of responding quickly.

Other microturbine technology needs, both present and future, are the standardization of interconnection and communications so that many units, even from different manufacturers, can be operated together in parallel. Power converters could benefit immediately from the introduction of silicon carbide (SiC) switching devices into the market due to their ability to operate at higher temperatures and other attractive characteristics discussed below.

The following topical discussions provide examples of useful research and critically needed tests that ORNL strongly recommends in support of the emerging microturbine industry. The following discussions offer strong technical justification for believing that these projects, if pursued, would provide significant cost savings and benefits to the manufacturers of power electronics.

1. *Control of real and reactive power in grid connect or stand alone mode.*

At the present time, power electronic converters used on microturbines operate in either voltage mode or current mode. Voltage mode is used when the microturbine is in stand alone, not grid connected, and the voltage level is set by the user. Current mode is used when the microturbine is grid connected. The current level and the power factor (for some manufacturers) are set by the user. As discussed above, it is desirable that the power electronics be capable of controlling both the real and reactive power. Voltage regulation cannot be provided without this capability, and reactive power will circulate among generators in a small-stand alone grid unless the reactive power is controlled.

One approach to resolving this issue would be to develop an enhancement to an existing power converter system so that it is capable of controlling voltage magnitude and phase angle (real and reactive power) in both the stand-alone mode and the grid connected mode. A project could be designed to evaluate the ability of the power-conditioning unit to enable real and reactive power sharing and to perform voltage regulation. The goal would be for voltage regulation to be provided in both the stand-alone and gridconnected modes. A project could be designed to test a power conditioning unit installed on a microturbine that has connections to both a local load and the distribution system. The work would evaluate, verify, and characterize the ability to perform voltage regulation under both operating modes.

2. *Seamless transition from grid-connect to stand-alone mode in subcycles*

In grid-connect mode, the utility interface works in current control, but in stand-alone mode it works in voltage control. To switch from one mode to the other, the microturbine and the interface could be shut down and then started in the other mode. This causes an undesirable voltage outage or brief fluctuation. A better alternative is a seamless transition without interrupting voltage production. Many manufacturers 26 offer "seamless" transition in their products but, in reality, it occurs in a few cycles causing voltage fluctuations. In some products, the transitions range from a few seconds to a few minutes.

The goal should be that the seamless transition from grid-connect to stand-alone modes and vice versa occur in a short, sub-cycle time. For this reason, research is required to come up with an algorithm to decrease the seamless transition time with minimum voltage distortion.

One approach would be to develop an enhancement to the existing Bowman power-conditioning unit so that it is capable of synchronizing frequency with the distribution system and changing control modes seamlessly. A project could be designed to test the device on a group of power sources and loads with a distribution system connection. The work would evaluate, verify, and characterize the ability to do a seamless transfer.

3. *Ganging microturbines with other microturbines and different energy sources:*

In many applications, microturbines in the field will be required to work in parallel with other microturbines or different energy sources such as fuel cells, diesel generators, photovoltaic systems, batteries, etc. Presently, only a few commercially available utility interfaces for microturbines, such as Capstone units, have a feature for connecting several microturbines together (other energy sources not included), and these mostly require extra accessories. However, this feature should be a standard built-in function of a general utility interface, and it should also include the ability to be ganged with other energy sources.

Three approaches are possible for the ganging of different energy sources. In the first approach, the output voltage of each energy source is converted to dc or high frequency ac. Then, these outputs are connected in series or parallel as required and converted to 60Hz with a dc link converter to be connected with the grid and/or to feed loads.

In the second approach, the output voltages of each energy sources are converted to dc and are connected to a multilevel inverter again to be connected to feed loads or directly to the utility grid.

In the third approach, each microturbine connects to the grid separately from the others relying wholly on their grid- or load-connect interface electronics. However, this is inconsistent with the approach we envision to enable the supply of ancillary services. Further enhancements are desirable where grid and/or load connection is coordinated instead of relying on multiple independent connections.

More research is required to evaluate the pros and cons of these approaches and to investigate the possibility of alternative solutions for ganging microturbines and other distributed energy resources.

Additional insight into solutions for the above issues can be gained by leveraging on other research being performed at ORNL on several different power electronics applications. The results of the following projects can be directly or indirectly applied to the microturbine power converters:

1. *Integrating numerous solid oxide fuel cell modules:* This project focuses on ways to connect fuel cell modules for higher power generation.

2. *Military generator sets:* This project is for the analysis, design, and building of mobile military generator sets for the Army.

3. *Hybrid electric vehicles (HEV):* Several projects are ongoing in the research and development of new power converter topologies, novel electrical machines, and their control. This also includes a system-level study of the benefits of SiC-based power devices on hybrid electric vehicle applications.

The experience gained from all these projects will be useful in microturbine applications. For example, the first project will provide answers to the similar problem of paralleling microturbines with each other and with other distributed generators. Additionally, experience from military generator sets and HEV power converters and electrical machines will be directly applicable to microturbine systems.

4. *Ability to interface with energy storage devices.*

The existing power converters from various manufacturers are capable of accepting a constant dc input. However, the voltage from many dc sources is variable such as from a fuel cell or an energy storage device (e.g., battery or ultra-capacitor). A dc-to-dc converter is needed. If the energy storage capability was available, the power converter could provide very high, short-duration power to start motors and supply fault current. This would make small systems much more viable and practical when operating in a stand-alone mode.

One approach to developing a dc interface would be to enhance an existing power converter so that it is capable of accepting a *variable* dc voltage input. A project could be designed to test the device on a group of power sources and loads to evaluate the ability to start motors, provide fault current, and regulate voltage during normal system operation.

5.   *Silicon Carbide switching devices*

Recently, silicon carbide- (SiC-) based power devices have been drawing increasing attention because of their superior characteristics compared with silicon- (Si-) based power devices. SiC-based power devices possess the features of high voltage, high power, high frequency, and high temperature operation in a smaller package. A SiC-based power converter would have the benefits of reduced losses, higher efficiency, up to 2/3 reduction in the heatsink size, smaller passive components, and less susceptibility to extreme ambient heat. Microturbine power converters would certainly benefit from utilizing SiC-power devices. The response from Ballard regarding what they would like to see from the Government or industry that would be of help to them was, *"Funding for High-voltage/Current Silicon Carbide diodes."* This response provides strong confirmation of the importance of SiC in this industry.

Further work is required to demonstrate the system-level benefits of SiC devices on microturbine power converter applications.

Index

Printed in the United States
by Baker & Taylor Publisher Services